Adenovirus Methods and Protocols, Second Edition
VOLUME 1

METHODS IN MOLECULAR MEDICINE™

John M. Walker, SERIES EDITOR

METHODS IN MOLECULAR MEDICINE™

Adenovirus Methods and Protocols

SECOND EDITION

*Volume 1: Adenoviruses, Ad Vectors,
Quantitation, and Animal Models*

Edited by

William S. M. Wold
Ann E. Tollefson

*Department of Molecular Microbiology and Immunology
Saint Louis University School of Medicine
St. Louis, Missouri*

HUMANA PRESS ✳ TOTOWA, NEW JERSEY

© 2007 Humana Press Inc.
999 Riverview Drive, Suite 208
Totowa, New Jersey 07512

humanapress.com

This publication is printed on acid-free paper. ∞
ANSI Z39.48-1984 (American Standards Institute)

Permanence of Paper for Printed Library Materials.

Cover illustration: Chapter 6, vol. 1, "Production and Release Testing of Ovine Atadenovirus Vectors," by Gerald W. Both, Fiona Cameron, Anne Collins, Linda J. Lockett, and Jan Shaw.

Production Editor: Amy Thau

Cover design by Donna Niethe

For additional copies, pricing for bulk purchases, and/or information about other Humana titles, contact Humana at the above address or at any of the following numbers: Tel.: 973-256-1699; Fax: 973-256-8341; E-mail: orders@humanapr.com; or visit our Website: www.humanapress.com

Photocopy Authorization Policy:
Authorization to photocopy items for internal or personal use, or the internal or personal use of specific clients, is granted by Humana Press Inc., provided that the base fee of US $30.00 per copy is paid directly to the Copyright Clearance Center at 222 Rosewood Drive, Danvers, MA 01923. For those organizations that have been granted a photocopy license from the CCC, a separate system of payment has been arranged and is acceptable to Humana Press Inc. The fee code for users of the Transactional Reporting Service is: [978-1-58829-598-9 • 1-58829-598-2/07 $30.00].

Printed in the United States of America. 10 9 8 7 6 5 4 3 2 1

eISBN: 1-59745-166-5

ISSN: 1543-1894

ISBN13: 978-1-58829-598-9

Library of Congress Cataloging-in-Publication Data

Adenovirus methods and protocols / edited by William S.M. Wold, Ann E. Tollefson. -- 2nd ed.
v. ; cm. -- (Methods in molecular medicine, ISSN 1543-1894 ; 130-131)
Includes bibliographical references and indexes.
Contents: v. 1. Adenoviruses, ad vectors, quantitation, and animal models -- v. 2. Ad proteins, RNA, lifecycle, host interactions, and phylogenetics.
ISBN 1-58829-598-2 (v. 1 : alk. paper) -- ISBN 1-58829-901-5 (v. 2 : alk. paper)
1. Adenoviruses--Laboratory manuals. 2. Molecular biology--Laboratory manuals. I. Wold, William S. M. II. Tollefson, Ann E. III. Series.
[DNLM: 1. Adenoviridae--Laboratory Manuals. W1 ME9616JM v.130-131 2007 / QW 25 A232 2007]
QR396.A336 2007
579.2'443--dc22
                    2006012284

Preface

Since their discovery in 1954, adenoviruses (Ads) have become premier and prolific models for studying virology, as well as molecular and cellular biology. Ads are benign to the researcher, easy to grow and manipulate, stable, versatile, and extremely interesting. In recent years, Ads have become ubiquitous in vector technology and in experimental gene therapy research.

The *Adenovirus Methods and Protocols* volumes are designed to help new researchers to conduct studies involving Ads and to help established researchers to branch into new areas. The chapters, which are written by prominent investigators, provide a brief, general introduction to a topic, followed by tried and true step-by-step methods pertinent to the subject. We thank returning contributors for their updated and new chapters and welcome new contributors who have expanded the content of this book.

The initial chapters of *Adenovirus Methods and Protocols, Second Edition, Volume 1: Adenoviruses, Ad Vectors, Quantitation, and Animal Models*, address several techniques for the construction of Ads for use as vectors and for basic research. Topics include deletion mutants, capsid modifications, insertions, and gene replacements in both human Ads and murine, bovine, and ovine Ads. There is detailed description of the quality-control testing of vectors. Two chapters address Ad vectors and apoptosis. Additional chapters describe methods for determination of virus titers, sensitive assays for the presence of viral DNA in samples, and quantitation of infection by detection of the expression of adenoviral proteins. Other new chapters address the development of new, permissive immunocompetent animal models (cotton rat and Syrian hamster) as Ads progress toward clinical applications.

We thank contributors for sharing their secrets, John Walker for his patience, and especially Dawn Schwartz, without whose expert assistance this work would not have been possible.

William S. M. Wold
Ann E. Tollefson

Contents

CONTENTS OF THE COMPANION VOLUME
Volume 2: Ad Proteins, RNA, Lifecycle, Host Interactions, and Phylogenetics

Contributors

G. ERIC BLAIR, PhD • *School of Biochemistry and Molecular Biology, University of Leeds, Leeds, United Kingdom*

GERALD W. BOTH, PhD • *CSIRO Molecular and Health Technologies, North Ryde, New South Wales, Australia*

GRAHAM BOTTLEY, MSc, PhD • *School of Biochemistry and Molecular Biology, University of Leeds, Leeds, United Kingdom*

JULIE BOYER, PhD • *Department of Genetic Medicine, Weill Medical College of Cornell University, New York, NY*

FIONA CAMERON, PhD • *CSIRO Molecular and Health Technologies, North Ryde, New South Wales, Australia*

ANGELA N. CAUTHEN, PhD • *Department of Natural Sciences, Clayton College and State University, Morrow, GA*

GOVINDASWAMY CHINNADURAI, PhD • *Institute for Molecular Virology, Saint Louis University School of Medicine, St. Louis, MO*

ANNE COLLINS, PhD • *Virient Party Ltd., Thebarton, South Australia, Australia*

THOMAS DOBNER, PhD • *Institut fuer Medizinische Mikrobiologie und Hygiene, Universitaet Regensburg, Regensburg, Germany*

KONSTANTIN DORONIN, PhD • *Department of Molecular Microbiology and Immunology, Saint Louis University School of Medicine, St. Louis, MO*

ERIK FALCK-PEDERSEN, PhD • *Department of Microbiology and Immunology, Weill Medical College of Cornell University, New York, NY*

JASON GALL, PhD • *Department of Microbiology and Immunology, Weill Medical College of Cornell University, New York, NY*

SHANTHI GANESH, MS • *Virotherapy, Cell Genesys, Inc., South San Francisco, CA*

C. T. GARNETT, PhD • *Laboratory of Tumor Immunology and Biology, National Cancer Institute, National Institutes of Health, Bethesda, MD*

LINDA R. GOODING, PhD • *Department of Microbiology and Immunology, Emory University School of Medicine, Atlanta, GA*

PETER GROITL, MSc • *Institut fuer Medizinische Mikrobiologie und Hygiene, Universitaet Regensburg, Regensburg, Germany*

LYNDA HAWKINS, PhD • *Diabetes/Metabolic Research, Novartis Institutes for BioMedical Research, Inc., Cambridge, MA*

JOHN R. HOLT, BSc, PhD • *School of Biochemistry and Molecular Biology, University of Leeds, Leeds, United Kingdom*

NEERAJA IDAMAKANTI, DVM, MS • *Department of Technology, Neotropix, Inc., Malvern, PA*

NICOLA J. JAMES, BSc, PhD • *School of Biochemistry and Molecular Biology, University of Leeds, Leeds, United Kingdom*

GARY KETNER, PhD • *Department of Molecular Microbiology and Immunology, Bloomberg School of Public Health, Johns Hopkins University, Baltimore, MD*

PETER KRAJCSI, PhD • *Research and Development Division, Solvo Biotechnology, Budaors, Hungary*

MOHAN KUPPUSWAMY, PhD • *Department of Molecular Microbiology and Immunology, Saint Louis University School of Medicine, St. Louis, MO*

DREW L. LICHTENSTEIN, PhD • *VirRx, Inc., St. Louis, MO*

LINDA J. LOCKETT, PhD • *CSIRO Molecular and Health Technologies, North Ryde, New South Wales, Australia*

VIVIEN MAUTNER, PhD • *CR UK Institute for Cancer Studies, University of Birmingham, Birmingham, United Kingdom*

CHING-I PAO, PhD • *Department of Microbiology and Immunology, Emory University, Atlanta, GA*

ELIZABETH PERRON, BS • *Laboratory of Gene Therapy, ENH-Research Institute and Department of Medicine, Evanston Hospital, Evanston, IL*

AMANDA PISTER, BS • *Laboratory of Gene Therapy, ENH-Research Institute and Department of Medicine, Evanston Hospital, Evanston, IL*

P. SESHIDHAR REDDY, DVM, PhD • *Department of Technology, Neotropix, Inc., Malvern, PA*

JOHN SCHOGGINS, BS • *Department of Microbiology and Immunology, Weill Medical College of Cornell University, New York, NY*

PREM SETH, PhD • *Evanston Hospital, Evanston, IL*

ELENA V. SHASHKOVA, PhD • *VirRx, Inc., St. Louis, MO*

JAN SHAW, PhD • *CSIRO Molecular and Health Technologies, North Ryde, New South Wales, Australia*

JACQUELINE F. SPENCER, BA, CVT • *Department of Molecular Microbiology and Immunology, Saint Louis University School of Medicine, St. Louis, MO*

KATHERINE R. SPINDLER, PhD • *Department of Microbiology and Immunology, University of Michigan Medical School, Ann Arbor, MI*

THIRUGNANA SUBRAMANIAN, PhD • *Institute for Molecular Virology, Saint Louis University School of Medicine, St. Louis, MO*

MARIA A. THOMAS, PhD • *Department of Molecular Microbiology and Immunology, Saint Louis University School of Medicine, St. Louis, MO*

JUN TIAN, MD • *Laboratory of Gene Therapy, ENH-Research Institute and Department of Medicine, Evanston Hospital, Evanston, IL*

SURESH K. TIKOO, DVM, PhD • *Vectored Vaccine Program VIDO, University of Saskatchewan, Saskatoon, Saskatchewan, Canada*

ANN E. TOLLEFSON, PhD • *Department of Molecular Microbiology and Immunology, Saint Louis University School of Medicine, St. Louis, MO*

KAROLY TOTH, DVM • *Department of Molecular Microbiology and Immunology, Saint Louis University School of Medicine, St. Louis, MO*

ZHEN-GUO WANG, PhD • *Laboratory of Gene Therapy, ENH-Research Institute and Department of Medicine, Evanston Hospital, Evanston, IL*

AMANDA R. WELTON, BS • *Department of Microbiology and Immunology, University of Michigan Medical School, Ann Arbor, MI*

WILLIAM S. M. WOLD, PhD • *Department of Molecular Microbiology and Immunology, Saint Louis University School of Medicine, St. Louis, MO*

QIAOHUA WU, PhD • *CBDS, Defence Research and Development, Ottowa, Alberta, Canada*

M. BEHZAD ZAFAR, MD • *Laboratory of Gene Therapy, ENH-Research Institute and Department of Medicine, Evanston Hospital, Evanston, IL*

ALEXANDER N. ZAKHARTCHOUK, DVM, PhD • *Vectored Vaccine Program VIDO, University of Saskatchewan, Saskatoon, Saskatchewan, Canada*

1

Manipulation of Early Region 4

Julie Boyer and Gary Ketner

Summary

Adenovirus early region 4 (E4) regulates processes in infected cells that include viral late gene expression, nonhomologous end joining, responses to DNA damage, and apoptosis. E4 is essential for viral growth in most cell lines. In this chapter, the current knowledge of the functions of six E4 products is summarized briefly. Protocols are presented for manipulation of E4, incorporation of E4 mutations into the viral genome, and growth of E4 mutants on complementing cell lines. A compilation of the described E4-complementing cell lines is included.

Key Words: Complementing cell line; DNA–protein complex; overlap recombination; ligation; bacterial recombination; mutant screening; adenovirus; early region 4; mutant; mutant construction; complementation.

1. Introduction

Early region 4 (E4) (**Fig. 1A**) occupies about 3000 base pairs at the right end of the human adenovirus genome. The sequences of E4 and E4 cDNAs indicate that E4 encodes seven polypeptides (*1–3*). E4 is essential for viral growth, and large E4 deletion mutations render mutant viruses defective for growth on normal host cells. Analyses of E4 mutants have implicated individual E4 products in a variety of processes that occur in infected cells, including viral early and late gene expression, DNA replication, the shut-off of host-cell protein synthesis, inhibition of nonhomologous end joining (NHEJ), transformation, and the ability to stimulate replication of adeno-associated virus. The biochemical functions of several of the E4 products are emerging. Although E4 is conserved among human adenoviruses, animal adenoviruses have divergent E4 regions, and some have no genes identifiably homologous to human E4 genes at all. How these viruses accomplish tasks carried out by human E4 genes is unclear. Current knowledge of the functions of E4 products is as follows:

From: *Methods in Molecular Medicine, Vol. 130:*
Adenovirus Methods and Protocols, Second Edition, vol. 1:
Adenoviruses, Ad Vectors, Quantitation, and Animal Models
Edited by: W. S. M. Wold and A. E. Tollefson © Humana Press Inc., Totowa, NJ

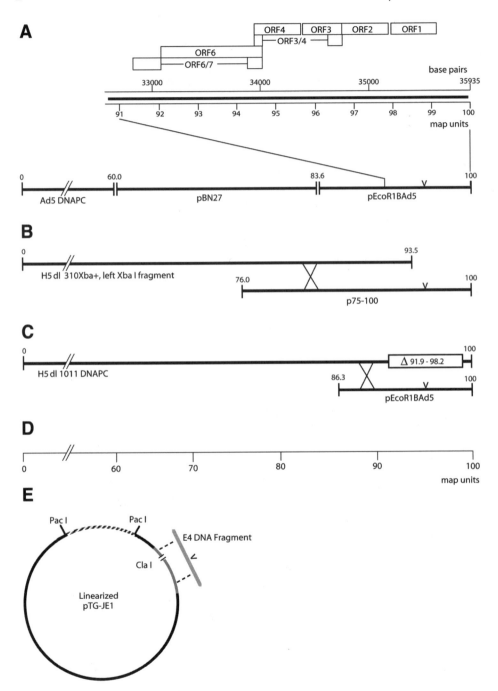

Fig. 1.

- Open reading frame (ORF)1: ORF1 mutants of adenovirus type 5 (Ad5) are viable in standard host cells and exhibit no mutant phenotypes. Uniquely, ORF1 of adenovirus 9 (Ad9) is responsible for the induction of mammary tumors in mice by that virus, and Ad9 E4 ORF1 alone can transform cells in culture *(4)*. Ad9 ORF1 stimulates phosphatidylinositol 3-kinase (PI3K) and downstream events in infected cells, an activity that is dependent upon its ability to bind to cellular proteins of the PDZ family *(5,6)*. Stimulation of PI3K mimics the activity of some growth factors, and this activity is presumably functionally important in tumorigenesis.
- ORF2: ORF2 mutants of Ad5 are viable and have no known mutant phenotypes in cultured cells. The ORF2 product is present in infected cells *(7)*, but there is no evidence implicating ORF2 in events in infected cells.
- ORF3 and ORF6: The products of ORFs 3 and 6 are required for viral late gene expression and thus for viral growth. They also stimulate viral DNA replication and prevent end-to-end concatenation of intracellular viral DNA by inhibiting NHEJ. Although they act by biochemically distinct methods (see below), ORFs 3 and 6 are genetically redundant: either product is sufficient for normal (ORF6) or near-normal (ORF3) function, but mutants lacking both are profoundly defective *(8–10)*. When constructing virus with mutations that simultaneously disrupt both ORFs, an E4-complementing cell line such as W162 *(11)* must be used for propagation. The viability of viruses expressing at least one or the other of these ORFs may be exploited when constructing mutants by the homologous recombination-based method described below. It should be noted that for all of the methods described, the slightly reduced ability of $ORF3^+/ORF6^-$ viruses to form plaques on normal hosts, while not absolutely demanding use of complementing cells, may make viruses of that genotype difficult to recover on other cell types.
- ORF6 forms a physical complex with the 55-kDa product of adenovirus early region 1b (E1b), and the complex mediates at least some of the functions of ORF6 *(12–14)*. It has been shown recently that the E4 ORF6/E1b 55k complex consti-

---

Fig. 1. (**A**) (*opposite page*) The organization of adenovirus early region 4 (E4). E4 open reading frames (ORFs) are indicated by open boxes above a scale that indicates the position in map units (0–100; 1 map unit is approx 360 bp) and nucleotide numbers. ORFs 1, 2, 3, 4, and 6 are colinear with the viral DNA; ORFs 3/4 and 6/7 are the result of inframe splicing of the E4 mRNA precursor. E4 is transcribed from right to left. (**B**) Reconstruction of an intact viral genome by ligation in vitro. (**C**) Reconstruction of an intact viral genome by overlap recombination. (**D**) Construction of an E4 mutant by recombination between a plasmid and an E4 deletion mutant. (**E**) Reconstruction of an E4 mutant by recombination in bacteria. In **C** and **D**, the X indicates a potential homologous recombination event; in **E**, potential homologous recombination events are indicated by dashed lines. In **B** through **E**, the position of a hypothetical E4 mutation is noted by a V. In **E**, the pTG-JE1 plasmid has been linearized by *Cla*I digestion. The grey portion of the plasmid flanked by *Pac*I sites contains the bacterial replicon and antibiotic selection markers. (**E** is modified from **ref. *41*.**)

tutes an ubiquitin E3 ligase that induces proteasome-dependent degradation of at least two cellular proteins, p53 and mre11 (a component of the cellular DNA damage-signaling pathway) *(15–18)*. p53 presumably is targeted to modulate apoptosis; degradation of mre11 prevents genome concatenation and abrogates activation of the DNA damage-signaling pathways in infected cells *(19)*. It is not yet known whether the ubiquitin E3 ligase activity of the complex is involved in stimulation of late gene expression. Other activities of the ORF6 product or the E4 ORF6/E1b 55k complex include modulation of p53 transcriptional activity by direct binding *(20)*, shut-off of host mRNA transport *(21)*, and modulation of alternative splicing in the adenoviral major late transcriptional unit *(22)*.

The activity of the E4 ORF3 protein seems to be mediated by its ability to rearrange the architecture of the infected cell nucleus. The ORF3 product is sufficient to disrupt intranuclear structures known as ND10 or PODs *(23)*, and the ORF3 product mislocalizes the mre11 protein *(17,24)*, presumably accounting for inhibition of concatenation. A unique role in late gene expression has not been identified, and the biochemical mechanism of ORF3's rearrangement of nuclear structure is not understood.

- ORF4: ORF4 mutants are viable in standard hosts. E4 ORF4 interacts with and modulates the activity of the cellular protein phosphatase 2A (PP2A) *(25)*, inducing hypophosphorylation of a variety of proteins in infected cells. Its several physiological effects all seem to be to the result of this interaction. For example, ORF4 depresses transcription from at least some promoters that contain CREB protein binding sites, including the adenovirus E2 and E4 promoters *(26,27)*, because of an ORF4-induced hypophosphorylation of cFos that results in a reduction in cJunB transcription and consequently JunB levels and AP-1 activity *(28)*. As a result of its effects on E2 transcription and, perhaps, on protein phosphorylation, ORF4 downregulates viral DNA synthesis. The E4 ORF 3 and 6 products antagonize this activity by an unknown mechanism *(29)*. ORF4 also induces apoptosis dependent on the PP2A interaction *(30,31)*.
- ORF6/7: ORF6/7 mutants are viable in standard hosts. The ORF6/7 product binds to the transcription factor E2F and confers cooperativity on E2F binding at promoters, such as the E2 promoter, that contain appropriately positioned E2F binding sites *(32–34)*. As a consequence, ORF6/7 expression stimulates E2 transcription in infected cells *(35)*. In most host cell types, this effect requires the E1a 289R protein, which dissociates E2F from pRB and makes it available for binding by ORF6/7 *(36)*. ORF6/7 mutants are not obviously defective in viral DNA synthesis in cultured cells.

1.1. Complementing Cell Lines

E4 mutants lacking both ORF3 and ORF6 are defective for growth in normal adenovirus hosts and must be propagated on cell lines capable of supplying required E4 functions *in trans*. The most commonly used such cell line is

W162 *(11)*. The W162 cell line was constructed by introducing the **Ad5 *Eco*RI** B fragment (map units 86.3 to 100) into Vero cells as part of a plasmid carrying the selectable marker gpt. W162 cells contain intact E4 and supply the E4 functions required to support the growth of deletions lacking all identified E4 products. E4 is attached to its natural promoter in W162 cells, and it is presumed that E4 expression is off until induced by E1a expression from an infecting virus. Both W162 and Vero cells (which are of monkey origin but permissive for human adenoviruses) support plaque formation by wild-type adenovirus about three times less efficiently than do other adenovirus hosts, such as 293 cells. It is therefore important when comparing defective E4 mutants titrated on W162 cells to other viruses on a PFU basis that W162 titers be obtained for all stocks.

In addition to W162 cells, several 293-derived cell lines that complement mutants with lesions in both E1 and E4 have been isolated. These lines, developed primarily for use in construction of gene delivery vectors, make it possible to grow multiply defective mutants. Because of its role in inhibiting host cell protein synthesis, E4 is presumed to be toxic if constitutively expressed. E1a expression induces the E4 promoter, and regulable heterologous promoters therefore have been used to control E4 expression in each of the 293-derived complementing cell lines. E4 expression thus remains low until it is deliberately induced. The E4-complementing cell lines described so far are listed in **Table 1**.

1.2. Isolation of Mutants

1.2.1. Naturally Occurring Mutants

The first E4 mutants were isolated from a stock of adenovirus 2 that had accumulated a substantial proportion of deletion mutants, presumably as a result of many undiluted serial passages. All of the mutants lacking E4 sequences were defective and were originally cloned and propagated by the use of a *ts* helper virus. For characterization, deletion mutant particles were purified from the resulting mixed stocks by repeated CsCl density gradient centrifugation *(37)* (*see also* Chapter 2). The development of E4-complementing cell lines made the use of helper viruses unnecessary, and most of what is known of the phenotypes of E4 mutants has been learned using mutants constructed in vitro and propagated on W162 cells.

1.2.2. Construction of Mutants In Vitro

Most E4 mutants are made by in vitro manipulation of E4-containing plasmids and subsequent introduction of the mutations into an intact viral genome. The mutations themselves can be made by any of the large variety of standard techniques. Several methods have been used to incorporate E4 mutations into

Table 1
E4-Complementing Cell Lines

Cell line	Mutations complemented	Promoter/inducer (if applicable)	Ref.
W162	E4	E4/none	Weinberg and Ketner *(11)*
VK2-20; VK10-9	E1, pIX, E4	MMTV/ dexamethasone	Krougliak and Graham *(43)*
IGRP2	E1, E4	MMTV/ dexamethasone	Yeh et al. *(44)*
A2	E1, E4	Metallothionein/ metal ions	Brough et al. *(45)*
293-E4	E1, E4	α-Inhibin/cAMP	Wang et al. *(46)*
2V6.11	E1, E4	Insect ecdysone-responsive/ Ponasterone A	Mohammadi et al. *(47)*

MMTV, mouse mammary tumor virus; cAMP, cyclic adenosine monophosphate.

infectious virus, including assembly of intact viral genomes by in vitro ligation of a mutant E4 DNA fragment to one or more subgenomic viral DNA fragments, recombination between a mutant E4 DNA fragment and an overlapping subgenomic fragment (or fragments) of the viral genome in transfected cells in culture, recombination between mutant plasmids and genomic clones in bacteria, and recombination between a mutant E4 DNA fragment and an intact viral genome in culture.

1.2.2.1. LIGATION IN VITRO (FIG. 1B)

A variety of viral and plasmid DNA fragments have been used for construction of E4 mutants by ligation. Both two-fragment and three-fragment schemes have been used. In most cases, a large restriction fragment covering the left-hand portion of the genome, derived from virion DNA or from virion-derived DNA–protein complex (DNAPC) *(38)*, is ligated to smaller plasmid-derived fragments that make up the remainder of the genome. The ligated DNA is then introduced into appropriate cells by transfection. The virion DNA fragment can be purified; alternatively, an unpurified fragment can be used if the chance of reassembling parental virus is reduced by digesting the virion DNA with an enzyme that cuts the right-hand end of the genome repeatedly. Neither of these approaches is completely effective in eliminating plaques produced by virus with the genotype of the left-end donor, and plaques that contain the desired recombinant are usually mixed. Therefore, plaques must be screened to identify those that contain recombinants, and recombinants usually must be plaque-purified. Constructions expected to yield defective E4 mutants must be done in complementing cell lines.

1.2.2.2. OVERLAP RECOMBINATION IN TRANSFECTED CELLS (FIG. 1C)

For the construction of E4 mutants by overlap recombination, a subgenomic left-hand terminal fragment of the viral genome is transfected into cells along with a plasmid bearing a mutant version of E4. The mutant plasmid must share sequences with the left-hand genomic fragment so that homologous recombination in the region of overlap can regenerate a full-length viral genome. Considerations such as purification of the left-hand fragment, the necessity for screening plaques, and the use of complementing cell lines are the same as those described for the ligation protocol above.

1.2.2.3. RECOMBINATION IN BACTERIA (FIG. 1E)

The availability of whole adenoviral genomic clones in bacteria and yeast has enabled a range of efficient schemes for construction of adenovirus mutants. Most commonly applied to generation of E1 mutants *(39,40)*, bacterial recombination can also be used to generate mutations in other regions *(41)*. Assembly of mutants in bacteria offers the major advantage that the actual cloning step (as well as structural verification and sequencing) takes place in fast-growing cells, eliminating the need for time-consuming purification by repeated plaquing.

1.2.2.4. RECOMBINATION WITH INTACT VIRAL DNA (FIG. 1D)

Many E4 mutants are viable in normal adenovirus host cell lines. Such mutants can be selected after co-transfection of normal hosts (such as 293 cells) with a mutant E4 DNA fragment and intact viral DNA prepared from a E4 deletion mutant that will not grow in those cells (e.g., H5*dl*1011) *(8)*. Because no plaque-forming virus is involved in this protocol, the background of unwanted plaques is much lower than with the other methods involving mutant construction in mammalian cells.

2. Materials

1. Tris-ethylene diamine tetraacetic acid (EDTA) (TE): 10 mM Tris-HCl, 1 mM EDTA, pH 8.1.
2. Buffered 8 M guanidinium hydrochloride solution: 8 M guanidinium hydrochloride, 50 mM Tris-HCl, pH 8.1.
3. Guanidinium hydrochloride/CsCl solutions: 4 M guanidinium hydrochloride containing either 3.03 M CsCl (51.0 g of CsCl per 100 mL of solution) or 4.54 M CsCl (76.5 g of CsCl per 100 mL of solution), buffered with 20 mM Tris-HCl. Final pH should be 8.1. Confirm the pH before use, as some lots of CsCl produce very acidic solutions unless neutralized.

4. PMSF: 100 m*M* phenylmethylsulfonyl fluoride in isopropanol. Store tightly capped at –20°C. PMSF is toxic.

5. HBS (*N*-2-hydroxyethylpiperazine-*N'*-2-ethanesulfonic acid [HEPES]-buffered saline): 8 g/L NaCl, 0.37 g/L KCl, 0.125 g/L $Na_2HPO_4 \cdot 2H_2O$, 1 g/L glucose, 5 g/L HEPES, final pH 7.05. The pH of the HBS is critical for the success of transfections.

6. Herring sperm DNA: dissolve herring sperm DNA in TE at approx 5 mg/mL. Sonicate on ice in 30-s bursts with an immersed probe until the viscosity no longer changes appreciably with each burst (6–10 bursts). Determine the concentration of the solution spectrophotometrically and adjust to 2.5 mg/mL with TE.

7. Plasmids (*see also* **Table 2**):
   a. pBN27 *(8)*: pBR322 containing Ad5 DNA from map unit (mu) 60 (*Bam*HI) to mu 86.4 (*Nde*I). One map unit corresponds to 1% of the viral genome, or approx 360 base pairs.
   b. pEcoRIBAd5 *(42)*: pBR322 containing Ad5 mu 83.6 (*Eco*RI) to mu 100 (right genomic end).
   c. pTG-JE1 *(41)*, a bacterial plasmid containing an entire adenoviral genome with a deletion mutation in E4 ORF6 and a *Cla*I linker inserted in E4 ORF3.

8. Viruses:
   a. Ad5.
   b. H5*dl*1011 *(8)*: a defective E4 mutant that lacks Ad5 nt 33091 to 35353.
   c. H5*dl*310Xba+ *(9)* carries a single *Xba*I site at mu 93.5.

9. Bacteria:
   a. *E. coli* XL-1 (Stratagene).
   b. *E. coli* BJ5183 (Stratagene).

10. Sucrose solutions: 5%, 20%, and 60% sucrose solutions prepared in 1 *M* NaCl, 1 m*M* EDTA, 10 m*M* Tris-HCl, pH 8.1.

11. 200X IGEPAL (Sigma catalog number I-3021): 10% IGEPAL CA-630 solution in water.

12. Phosphate-free medium: minimum essential medium (MEM) formulated without sodium phosphate can be purchased from many media suppliers. For labeling viral DNA, supplement medium with 2% fetal bovine serum (FBS) (dialysis is not necessary) and 40 µCi/mL carrier-free ^{32}P orthophosphate.

13. Lysis buffer: 0.6% sodium dodecyl sulfate (SDS), 200 µg proteinase K, 10 m*M* Tris-HCl, pH 8.1, 1 m*M* EDTA.

14. 5 *M* Ammonium acetate.

15. 1,1,2-Trichlorotrifluoroethane (Sigma, ca. no. T5271).

16. 25% (w/v) glucose in HBS; filter sterilize.

17. Agar medium: MEM without phenol red, supplemented with 2% FBS, penicillin, and streptomycin, and solidified with 0.9% Difco Bacto-agar. Prepare and filter-sterilize 2X MEM without sodium bicarbonate, glutamine, and phenol red. Add those components, supplement with FBS, penicillin, streptomycin, and glutamine; store tightly closed at 4°C. Note that this medium is 2X, and supplements should be added accordingly. Prepare and autoclave 1.8% agar in water.

Table 2
Construction of E4 Mutants by Ligation and Overlap Recombination

Virion DNA fragment	Cloned fragment(s)	Reference
Ligation:		
0–60.0 (*Bam*HI) from Ad5 DNA PC	60.0 (*Bam*HI) to 83.6 (*Eco*RI) from pBN27; 83.6 (*Eco*RI) to 100 from pEcoRIBAd5	Bridge and Ketner *(8)*
0–76.0 (*Eco*RI) of H5*wt*300	76.0 (*Eco*RI) to 100 from p75-100	Huang and Hearing *(9)*
0–76.0 (*Eco*RI) of H5*dl*343; 76.0 (*Eco*RI) to 100 from H5*dl*356	None	Marton et al. *(34)*
Overlap recombination:		
0–93.5 (*Xba*I) of H5*dl*310Xba$^+$	76.0 (*Eco*RI)–100 from p75-100	Huang and Hearing *(9)*
0–79.6 (*Xba*I) from H5*wt*300	60.0 (*Bam*HI) to 100; 79.6 (*Xba*I) to 84.8 (*Xba*I) deleted in construction	Halbert et al. *(10)*
None	0 to 76.0 (*Eco*RI), 73.2 (*Hin*dIII) to 89.0 (*Hin*dIII), 83.6 (*Eco* RI) to 100	Hemstrom et al. *(48)*

Genome positions are given in map units (mu); 1 mu is approx 360 bp.

To prepare agar overlays, melt the agar in a steam bath or microwave oven, mix with an equal volume of supplemented 2X MEM, equilibrate to 48°C, and use to overlay plates.

3. Methods

3.1. Preparation of Adenovirus DNAPC (38)

1. Dialyze CsCl density gradient purified virus particles (protocol appears in Chapter 2) against TE. Highly concentrated virus suspensions sometimes precipitate during dialysis; this does not affect the procedure.
2. Mix equal volumes of virus suspension and buffered 8 *M* guanidinium hydrochloride (GuHCl) solution. Immediately add PMSF to 1 m*M* final concentration.

Incubate on ice for 5 min. The turbidity of the suspension will decrease as the virus particles are dissociated.

3. To the disrupted virus, add 4 *M* GuHCl, 4.54 *M* CsCl, 20 m*M* Tris-HCl, pH 8.1. The volume should be four times the original volume of the virus suspension.

4. Adjust to a convenient volume with 4 *M* GuHCl, 3.03 *M* CsCl, 20 m*M* Tris-HCl, pH 8.1.

5. Centrifuge the solution overnight at 177,000*g* and 15°C in a Sorvall TV865 rotor or equivalent.

6. Collect 0.25- to 0.5-mL fractions and locate fractions containing DNA by UV spectrophotometry.

7. Pool DNA-containing fractions and dialyze 4 h vs TE, 1 *M* NaCl, then overnight vs two changes of TE. Store the DNAPC in a plastic tube, as the terminal protein adheres to glass. The yield should be approx 10 µg of DNAPC per 10-cm Petri dish of infected cells.

3.2. Construction of E4 Mutants by Ligation

1. Introduce the desired E4 mutation into the plasmid pEcoRIBAd5.

2. Prepare Ad5 or H5*dl*1011 DNAPC. The use of H5*dl*1011 reduces the background of plaques containing parental virus when constructing viable E4 mutants in noncomplementing cell lines.

3. Prepare pBN27 and mutant pEcoRIBAd5 DNAs by CsCl–ethidium bromide density gradient centrifugation, Qiagen column, or similar method.

4. Digest 9 µg of DNAPC with *Bam*HI and *Eco*RI in a microfuge tube. This yields a large left-end fragment (0–60.0 mu) and cleaves the right end into three pieces.

5. Precipitate the digested DNA by the addition of 2.5 volumes of ethanol directly to the digestion reaction. Chill on ice for 15 min, centrifuge for 15 min at 4°C in a microfuge, and remove the supernatant with a micropipette. Invert the tube and air-dry the pellet for 15 min at room temperature.

6. Dissolve the dried pellet in a small volume of TE (*see* **Note 1**).

7. Digest 15 µg of pBN27 with *Bam*HI and *Eco*RI and the mutant E4 plasmid with *Eco*RI. Precipitate the digested DNAs as described in **step 5** and dissolve in small volumes of TE.

8. Add the digested plasmid DNAs to the dissolved digested DNAPC (*see* **Note 1**). Add concentrated ligase buffer to 1X final concentration and add 5–7 U of T4 DNA ligase. Total volume should be 70 µL.

9. Incubate overnight at 4°C.

10. Transfect the ligated DNA into W162 or other complementing cells or, if the mutant is likely to be viable, into 293 cells. Use 2.5 µg of ligated DNA and 7.5 µg of herring sperm DNA per transfected 60-mm dish.

11. Pick plaques and screen for mutants as described in **Subheading 3.7**.

3.3. Construction of E4 Mutants by Overlap Recombination

1. Introduce the desired E4 mutation into the plasmid pEcoRIBAd5.

2. Digest the mutant plasmid with *Eco*RI.

3. Digest purified H5*dl*310Xba+ virion DNA with *Xba*I.
4. Purify the large (left-hand) *Xba*I fragment.
   a. Load 10–50 μg of digested DNA on a linear 5–20% sucrose gradient built over a 60% sucrose cushion in an SW40 ultracentrifuge tube. (All sucrose solutions are in TE, with 1 *M* NaCl.)
   b. Centrifuge the gradients at 82,300*g* for 18 h at room temperature.
   c. Fractionate the gradient into 0.5- to 1-mL fractions.
   d. Identify the fractions containing the large *Xba*I fragment by analyzing small portions of each fraction on an agarose gel.
   e. Pool and briefly (2 h) dialyze positive fractions.
   f. Ethanol-precipitate the DNA.
5. Transfect appropriate cells. Use 1 μg of large fragment and 9 μg of digested plasmid DNA per 60-mm dish.
6. Pick plaques and screen for mutants as described in **Subheading 3.7.**

3.4. Construction of E4 Mutants by Recombination in Bacteria (see ref. 41)

1. Introduce the desired mutation into an E4 plasmid such as pEcoRIBAd5. For use with the specific recipient indicated below, which was designed for use in creating ORF3 mutations *(41)*, the mutant plasmid must contain E4 ORF3 (wild-type or mutant) and adenoviral sequences that flank both E4 ORF3 and the mutation to be introduced. Recombination occurs in these flanking sequences.
2. Prepare plasmid pTG-JE1 DNA, linearize by digestion with *Cla*I, and treat with calf intestinal phosphatase to prevent recircularization after transformation. (Plasmid pTG-JE1 contains two *Cla*I sites: one in ORF3 and one in E1. However, the site in E1 is methylated in plasmids grown in *dam*+ bacteria and will not be cleaved.)
3. Excise the E4 region from the mutant plasmid. As indicated above, the fragment must contain adenovirus sequences flanking ORF3 and the mutation to be introduced into virus.
4. Co-transform *E. coli* BJ5183 with 500 ng of the mutant E4 DNA fragment and 100 ng of linearized pTG-JE1 by electroporation, selecting recombinants with ampicillin and streptomycin.
5. Extract plasmids from transformants and confirm the presence of the mutation by sequencing or restriction enzyme digestion (*see* **Note 2**).
6. Propagate and prepare the recombinant plasmids in *E. coli* XL-1.
7. Excise the mutant genome from the plasmid by *Pac*I digestion.
8. Introduce 10 μg of excised DNA into W162 cells by calcium phosphate transfection.
9. 8 d postinfection, harvest the cells and plaque-purify mutant virus once on W162 cells.

3.5. Construction of E4 Mutants by Recombination With an Intact Viral Genome

1. Introduce the desired E4 mutation into the plasmid pEcoRIBAd5.
2. Prepare H5*dl*1011 DNAPC.

3. Linearize the E4 mutant plasmid in adenovirus sequences outside of the H5*dl*1011 deletion by digestion with *Hpa*I (nt. 32002) or *Nde*I (nt. 31088). This results in a 1- to 2-kb region to the left of E4, where recombination can occur.
4. Combine 2 µg each of the digested plasmid DNA and H5*dl*1011 DNAPC (3.6:1 molar ratio of plasmid DNA to DNAPC). Add 6 µg sonicated herring sperm DNA, for 10 µg total DNA per transfection.
5. Transfect this mixture into 293 cells. This number of cells can be transfected by a procedure identical to that for W162 cells (*see* **Subheading 3.6.**), with the omission of the glucose boost step.
6. Pick plaques that arise and screen as in **Subheading 3.7.** (*see* **Note 3**).

3.6. Calcium Phosphate Transfection (31,32)

1. Prepare freshly confluent monolayers of W162 (for defective E4 mutants) or 293 (for mutants that do not require an E4 complementing line) cells in 60-mm tissue culture dishes. Slightly subconfluent monolayers are acceptable but can be fragile, and heavily confluent monolayers can be used with somewhat reduced efficiency.
2. Add 10 µg of plasmid, viral, and sonicated herring sperm DNA, in a minimum volume of TE or ligation buffer (less than 50 µL), to 0.95 mL of HBS in a plastic tube. The amounts of the various DNAs recommended for each procedure are noted above.
3. Add 50 µL of 2.5 *M* CaCl$_2$ and mix quickly.
4. Incubate 20–30 min at room temperature.
5. Without removing the medium, add 0.5 mL of the mixture to each of two 60-mm dishes containing cell monolayers. Gently agitate the dishes to distribute the DNA evenly.
6. Incubate 4–6 h at 37°C.
7. For W162 cells, boost with a 25% glucose shock (*see* **Note 4**).
   a. Remove the medium from a transfected dish.
   b. Gently add 2 mL of sterile 25% glucose directly to the center of the monolayer.
   c. Incubate for 4 min exactly, remove the glucose, and rinse the monolayer once with culture medium.
   d. Overlay with 5 mL of agar medium.
8. Incubate the dishes at 37°C, adding 2.5 mL of agar medium every third day. The second addition should contain neutral red.
9. When plaques arise (7–12 d), mark their locations on the outside of the plaquing dishes.
10. Pick plaques for screening by aspirating the agar and cell debris over and around a marked plaque into a Pasteur pipet.
11. Transfer the agar and cell debris into a vial containing 1 mL of culture medium. Store at –80°C.

3.7. Screening for Mutants

1. Prepare a high-titer ministock from each plaque.
   a. Inoculate monolayers of permissive cells growing in a 24-well tissue culture plate with the resuspended plaque, frozen and thawed three times to release virus from the cell debris.
   b. Inspect the wells daily, replacing the medium every third day; harvest the stocks when most or all of the cells in a well have detached from the plastic.
   c. Freeze and thaw the ministock three times.
2. Inoculate monolayers of permissive cells growing in 24-well tissue culture plates with 100 µL of the high-titer ministock.
3. Allow the virus to adsorb for 2 h at 37°C, then remove the inoculum and refill the well with 2 mL of medium.
4. 24 h after infection, remove the medium from the infected cells, rinse the monolayer once with phosphate-free medium, then add 1 mL of phosphate-free medium containing 2% FBS and 40 µCi/mL ^{32}P orthophosphate.
5. Incubate 6–24 h at 37°C.
6. Remove the labeling medium and gently rinse the monolayers twice with phosphate-buffered saline.
7. Lyse the labeled cells by adding 300 µL of lysis buffer (**Heading 2.**, **item 13**) directly to the wells.
8. Seal the plate with Parafilm and incubate at 37°C, 2 h to overnight.
9. Transfer the contents of each well to a microfuge tube, add 200 µL of 5 *M* ammonium acetate and mix well.
10. Add 1 mL isopropyl alcohol and mix well.
11. Centrifuge immediately for 5 min at full speed in a microcentrifuge.
12. Rinse the pellet twice in 70% ethanol.
13. Air-dry the pellet and resuspend the DNA in TE.
14. The labeled DNA can be analyzed by restriction digestion, agarose gel electrophoresis, and autoradiography for the presence of the mutation.
15. After identification of plaques containing the desired recombinants, use the ministock (**Subheading 3.7.**, **step 1**) for a second round of plaque purification. Before plaquing, add IGEPAL to 0.05% and extract the stock vigorously with 1/10 vol of 1,1,2-trichlorotrifluoroethane. This dissociates clumps of virus that otherwise make plaque purification extremely inefficient. Plaques should be purified one round beyond the point at which parental (Ad5 or H5*dl*1011) DNA can no longer be detected in any plaque.

4. Notes

1. DNAPC has a tendency to bind to the walls of tubes, etc., and so the digestion of the DNAPC and ligation should be carried out in the same tube if possible. To preserve the terminal protein, do not treat the digest with protease, phenol, or SDS.
2. Depending on the distance between the mutation and the site of linearization in ORF3, a variable proportion of the recombinants will have experienced cross-

overs between ORF3 and the mutation and will not incorporate the mutation. It is therefore most efficient to construct mutants that can easily be identified by restriction digestion rather than relying on sequencing.

3. Almost all plaques will contain both recombinant virus and H5*dl*1011, the latter complemented by the viable recombinant. When preparing minilysates for screening (*see* **Subheading 3.7.**), use cells that are nonpermissive for H5*dl*1011 to enrich the ministocks for the recombinant. However, labeled DNA should be prepared in W162 cells, which permit growth of H5*dl*1011 and thus maximize the sensitivity of the screening procedure for contaminating parental virus.

4. Timing of the glucose boost is important for good efficiency and monolayer survival; if multiple dishes are to be boosted as a group, stagger the addition of the glucose solution so that precise timing can be maintained. To simplify the manipulation of large numbers of dishes, dishes can be refilled with culture medium after the glucose boost and overlaid with agar later as a group.

References

1. Freyer, G. A., Katoh, Y., and Roberts, R. J. (1984) Characterization of the major mRNAs from adenovirus 2 early region 4 by cDNA cloning and sequencing. *Nucleic Acids Res.* **12,** 3503–3519.
2. Dix, I. and Leppard, K. N. (1992) Open reading frames 1 and 2 of adenovirus region E4 are conserved between human serotypes 2 and 5. *J. Gen. Virol.* **73,** 2975–2976.
3. Virtanen, A., Gilardi, P., Naslund, A., LeMoullec, J. M., Pettersson, U., and Perricaudet, M. (1984) mRNAs from human adenovirus 2 early region 4. *J. Virol.* **51,** 822–831.
4. Javier, R. T. (1994) Adenovirus type 9 E4 open reading frame 1 encodes a transforming protein required for the production of mammary tumors in rats. *J. Virol.* **68,** 3917–3924.
5. Frese, K. K., Lee, S. S., Thomas, D. L., Latorre, I. J., Weiss, R. S., Glaunsinger, B. A., and Javier, R. T. (2003) Selective PDZ protein-dependent stimulation of phosphatidylinositol 3-kinase by the adenovirus E4-ORF1 oncoprotein. *Oncogene* **22,** 710–721.
6. O'Shea, C., Klupsch, K., Choi, S., Bagus, B., Soria, C., Shen, J., McCormick, F., and Stokoe, D. (2005) Adenoviral proteins mimic nutrient/growth signals to activate the mTOR pathway for viral replication. *EMBO J.* **24,** 1211–1221.
7. Dix, I. and Leppard, K. N. (1995) Expression of adenovirus type 5 E4 Orf2 protein during lytic infection. *J. Gen. Virol.* **76,** 1051–1055.
8. Bridge, E. and Ketner, G. (1989) Redundant control of adenovirus late gene expression by early region 4. *J. Virol.* **63,** 631–638.
9. Huang, M. M. and Hearing, P. (1989) Adenovirus early region 4 encodes two gene products with redundant effects in lytic infection. *J. Virol.* **63,** 2605–2615.
10. Halbert, D. N., Cutt, J. R., and Shenk, T. (1985) Adenovirus early region 4 encodes functions required for efficient DNA replication, late gene expression, and host cell shutoff. *J. Virol.* **56,** 250–257.

11. Weinberg, D. H. and Ketner, G. (1983) A cell line that supports the growth of a defective early region 4 deletion mutant of human adenovirus type 2. *Proc. Natl. Acad. Sci. USA* **80,** 5383–5386.
12. Cutt, J. R., Shenk, T., and Hearing, P. (1987) Analysis of adenovirus early region 4-encoded polypeptides synthesized in productively infected cells. *J. Virol.* **61,** 543–552.
13. Bridge, E. and Ketner, G. (1990) Interaction of adenoviral E4 and E1b products in late gene expression. *Virology* **174,** 345–353.
14. Sarnow, P., Hearing, P., Anderson, C. W., Halbert, D. N., Shenk, T., and Levine, A. J. (1984) Adenovirus early region 1B 58,000-dalton tumor antigen is physically associated with an early region 4 25,000-dalton protein in productively infected cells. *J. Virol.* **49,** 692–700.
15. Querido, E., Blanchette, P., Yan, Q., Kamura, T., Morrison, M., Boivin, D., Kaelin, W. G., Conaway, R. C., Conaway, J. W., and Branton, P. E. (2001) Degradation of p53 by adenovirus E4orf6 and E1B55K proteins occurs via a novel mechanism involving a Cullin-containing complex. *Genes Dev.* **15,** 3104–3117.
16. Harada, J. N., Shevchenko, A., Pallas, D. C., and Berk, A. J. (2002) Analysis of the adenovirus E1B-55K-anchored proteome reveals its link to ubiquitination machinery. *J. Virol.* **76,** 9194–9206.
17. Stracker, T. H., Carson, C. T., and Weitzman, M. D. (2002) Adenovirus oncoproteins inactivate the Mre11-Rad50-NBS1 DNA repair complex. *Nature* **418,** 348–352.
18. Cathomen, T. and Weitzman, M. D. (2000) A functional complex of adenovirus proteins E1B-55kDa and E4orf6 is necessary to modulate the expression level of p53 but not its transcriptional activity. *J. Virol.* **74,** 11,407–11,412.
19. Carson, C. T., Schwartz, R. A., Stracker, T. H., Lilley, C. E., Lee, D. V., and Weitzman, M. D. (2003) The Mre11 complex is required for ATM activation and the G2/M checkpoint. *EMBO J.* **22,** 6610–6620.
20. Dobner, T., Horikoshi, N., Rubenwolf, S., and Shenk, T. (1996) Blockage by adenovirus E4orf6 of transcriptional activation by the p53 tumor suppressor. *Science* **272,** 1470–1473.
21. Flint, S. J. and Gonzalez, R. A. (2003) Regulation of mRNA production by the adenoviral E1B 55-kDa and E4 Orf6 proteins. *Curr. Top. Microbiol. Immunol.* **272,** 287–330.
22. Nordqvist, K., Ohman, K., and Akusjarvi, G. (1994) Human adenovirus encodes two proteins which have opposite effects on accumulation of alternatively spliced mRNAs. *Mol. Cell. Biol.* **14,** 437–445.
23. Carvalho, T., Seeler, J. S., Ohman, K., Jordan, P., Pettersson, U., Akusjarvi, G., Carmo-Fonseca, M., and Dejean, A. (1995) Targeting of adenovirus E1A and E4-ORF3 proteins to nuclear matrix-associated PML bodies. *J. Cell Biol.* **131,** 45–56.
24. Evans, J. D. and Hearing, P. (2005) Relocalization of the Mre11-Rad50-Nbs1 complex by the adenovirus E4 ORF3 protein is required for viral replication. *J. Virol.* **79,** 6207–6215.

25. Kleinberger, T. and Shenk, T. (1993) Adenovirus E4orf4 protein binds to protein phosphatase 2A, and the complex down regulates E1A-enhanced junB transcription. *J. Virol.* **67,** 7556–7560.

26. Bondesson, M., Ohman, K., Manervik, M., Fan, S., and Akusjarvi, G. (1996) Adenovirus E4 open reading frame 4 protein autoregulates E4 transcription by inhibiting E1A transactivation of the E4 promoter. *J. Virol.* **70,** 3844–3851.

27. Medghalchi, S., Padmanabhan, R., and Ketner, G. (1997) Early region 4 modulates adenovirus DNA replication by two genetically separable mechanisms. *Virology* **236,** 8–17.

28. Muller, U., Kleinberger, T., and Shenk, T. (1992) Adenovirus E4orf4 protein reduces phosphorylation of c-Fos and E1A proteins while simultaneously reducing the level of AP-1. *J. Virol.* **66,** 5867–5878.

29. Bridge, E., Medghalchi, S., Ubol, S., Leesong, M., and Ketner, G. (1993) Adenovirus early region 4 and viral DNA synthesis. *Virology* **193,** 794–801.

30. Shtrichman, R. and Kleinberger, T. (1998) Adenovirus type 5 E4 open reading frame 4 protein induces apoptosis in transformed cells. *J. Virol.* **72,** 2975–2982.

31. Marcellus, R. C., Lavoie, J. N., Boivin, D., Shore, G. C., Ketner, G., and Branton, P. E. (1998) The early region 4 orf4 protein of human adenovirus type 5 induces p53-independent cell death by apoptosis. *J. Virol.* **72,** 7144–7153.

32. Huang, M. M. and Hearing, P. (1989) The adenovirus early region 4 open reading frame 6/7 protein regulates the DNA binding activity of the cellular transcription factor, E2F, through a direct complex. *Genes Dev.* **3,** 1699–1710.

33. Neill, S. D., Hemstrom, C., Virtanen, A., and Nevins, J. R. (1990) An adenovirus E4 gene product trans-activates E2 transcription and stimulates stable E2F binding through a direct association with E2F. *Proc. Natl. Acad. Sci. USA* **87,** 2008–2012.

34. Marton, M. J., Baim, S. B., Ornelles, D. A., and Shenk, T. (1990) The adenovirus E4 17-kilodalton protein complexes with the cellular transcription factor E2F, altering its DNA-binding properties and stimulating E1A-independent accumulation of E2 mRNA. *J. Virol.* **64,** 2345–2359.

35. Swaminathan, S. and Thimmapaya, B. (1996) Transactivation of adenovirus E2-early promoter by E1A and E4 6/7 in the context of viral chromosome. *J. Mol. Biol.* **258,** 736–746.

36. Raychaudhuri, P., Bagchi, S., Neill, S. D., and Nevins, J. R. (1990) Activation of the E2F transcription factor in adenovirus-infected cells involves E1A-dependent stimulation of DNA-binding activity and induction of cooperative binding mediated by an E4 gene product. *J. Virol.* **64,** 2702–2710.

37. Challberg, S. S. and Ketner, G. (1981) Deletion mutants of adenovirus 2: isolation and initial characterization of virus carrying mutations near the right end of the viral genome. *Virology* **114,** 196–209.

38. Robinson, A. J. and Bellett, J. D. (1975) A circular DNA-protein complex adenoviruses and its possible role in DNA replication. *Cold Spring Harb. Symp. Quant. Biol.* **39,** 523–531.

39. He, T. C., Zhou, S., da Costa, L. T., Yu, J., Kinzler, K. W., and Vogelstein, B. (1998) A simplified system for generating recombinant adenoviruses. *Proc. Natl. Acad. Sci. USA* **95**, 2509–2514.

40. Chartier, C., Degryse, E., Gantzer, M., Dieterle, A., Pavirani, A., and Mehtali, M. (1996) Efficient generation of recombinant adenovirus vectors by homologous recombination in Escherichia coli. *J. Virol.* **70**, 4805–4810.

41. Evans, J. D. and Hearing, P. (2003) Distinct roles of the Adenovirus E4 ORF3 protein in viral DNA replication and inhibition of genome concatenation. *J. Virol.* **77**, 5295–5304.

42. Berkner, K. L. and Sharp, P. A. (1983) Generation of adenovirus by transfection of plasmids. *Nucleic Acids Res.* **11**, 6003–6020.

43. Krougliak, V. and Graham, F. L. (1995) Development of cell lines capable of complementing E1, E4, and protein IX defective adenovirus type 5 mutants. *Hum. Gene Ther.* **6**, 1575–1586.

44. Yeh, P., Dedieu, J. F., Orsini, C., Vigne, E., Denefle, P., and Perricaudet, M. (1996) Efficient dual transcomplementation of adenovirus E1 and E4 regions from a 293-derived cell line expressing a minimal E4 functional unit. *J. Virol.* **70**, 559–565.

45. Brough, D. E., Lizonova, A., Hsu, C., Kulesa, V. A., and Kovesdi, I. (1996) A gene transfer vector-cell line system for complete functional complementation of adenovirus early regions E1 and E4. *J. Virol.* **70**, 6497–6501.

46. Wang, Q., Jia, X. C., and Finer, M. H. (1995) A packaging cell line for propagation of recombinant adenovirus vectors containing two lethal gene-region deletions. *Gene Ther.* **2**, 775–783.

47. Mohammadi, E. S., Ketner, E. A., Johns, D. C., and Ketner, G. (2004) Expression of the adenovirus E4 34k oncoprotein inhibits repair of double strand breaks in the cellular genome of a 293-based inducible cell line. *Nucleic Acids Res.* **32**, 2652–2659.

48. Hemstrom, C., Virtanen, A., Bridge, E., Ketner, G., and Pettersson, U. (1991) Adenovirus E4-dependent activation of the early E2 promoter is insufficient to promote the early-to-late-phase transition. *J. Virol.* **65**, 1440–1449.

2

Isolation, Growth, and Purification of Defective Adenovirus Deletion Mutants

Gary Ketner and Julie Boyer

Summary

Defective adenovirus deletion mutants can be grown by complementation in the presence of helper viruses that supply essential functions missing in the deletion mutant. In general, the deletion mutant then must be separated physically from the helper for use in subsequent experiments. This chapter includes suggestions for selection of helper viruses, protocols for the production of stocks by complementation, and procedures for physical separation of deletion mutants from their helpers.

Key Words: Deletion mutant; complementation; complementation plaquing; density gradient centrifugation; radiolabeling; helper virus.

1. Introduction

Adenovirus mutants that lack essential genes must be grown by complementation, the products of the missing genes supplied by a source other than the viral genome. Two complementation methods are available for the growth of defective adenovirus mutants. For mutations in E1, E2, E4, or proteins IV (fiber), IVa2, IX, or the 23k late protease, complementing cell lines that contain segments of viral DNA and that can supply the missing viral products can be used to produce pure stocks of mutant particles *(1–11)*. This approach will probably be extended to other regions of the viral genome but may prove difficult to adapt to genes encoding abundant capsid proteins whose products are required in large amounts by the virus.

Alternatively, defective mutants can be grown as mixed stocks with a second "helper" virus that can supply *in trans* functions required by the mutant *(12)*. Providing that a mutant contains all of the *cis*-active elements required for viral growth and is large enough to be packaged into an adenoviral capsid,

From: *Methods in Molecular Medicine, Vol. 130:*
Adenovirus Methods and Protocols, Second Edition, vol. 1:
Adenoviruses Ad Vectors, Quantitation, and Animal Models
Edited by: W. S. M. Wold and A. E. Tollefson © Humana Press Inc., Totowa, NJ

there are in principle no restrictions on the DNA sequences that can be deleted from a mutant grown by complementation with helper virus. Further, because the helper virus replicates, even products needed in large amounts can be effectively supplied *in trans*. Extreme examples of defective adenoviruses propagated by complementation from a helper virus are provided by gene therapy vectors that lack nearly all viral sequences *(13,14)*. These helper-dependent or "gutless" gene transfer vectors were developed both to minimize antivector immune responses that restrict the duration of transgene expression from conventional vectors and to maximize capacity for accommodating transgenes. The size constraints imposed by some large genes such as dystrophin require that most of the adenoviral genome be deleted simply to make space for the therapeutic gene, and increased vector capacity also allows use of the endogenous promoter and control elements to naturally and precisely regulate the expression level of the therapeutic gene *(15)*.

Growth of mutants or gene therapy vectors by complementation with helper virus requires that, for most purposes, the mutant and helper be physically separated before use. This generally is done by CsCl equilibrium density gradient centrifugation. The following protocols were developed for propagation of defective mutants with modest deletions (10–20% of the viral genome), but are applicable to larger deletions and to substitution mutants with genome sizes that differ from that of wild-type virus.

1.1. Growth of Deletion Mutants as Mixed Stocks

Because no two viral mutants have identical growth characteristics, the composition of a mixed virus stock changes over time. In particular, a defective mutant grown in the presence of replication-competent helper virus tends to disappear from the stock because any cell infected by such a mutant alone yields no progeny, while cells singly infected by the helper produce a normal yield of virus particles. Several approaches can be used to minimize that tendency. First, if a mutant helper that requires complementation for its own growth can be used, and if the deletion mutant of interest complements the helper, only dually infected cells will produce particles. Alternatively, a helper virus with defective packaging signals can be used. Such viruses are packaged into virions with much lower efficiency than the virus with normal packaging signals or, as in the case of some gene therapy vectors, two intact packaging signals. The multiplicity of infection (MOI) used to produce stocks can be made high enough to ensure that virtually all infected cells contain both a helper virus and the defective mutant. Finally, seed stocks can be enriched for the deletion mutant by physical methods before use in preparing new stocks. The use of multiple approaches when possible maximizes the yield of mutant particles.

1.2. Selection of Helper Virus

Three criteria should be used in selecting helper virus:

1. If possible, the helper should be defective and require complementation by the mutant of interest for growth or should carry a defective packaging signal. For example, helpers carrying temperature-sensitive (*ts*) mutations in late genes have been used in the isolation of Ad2 E4 deletion mutants (*12*), and packaging-defective helpers have been used for complementing gutless vectors (*16–18*).
2. Because separation of the mutant and helper depends on differences in buoyant density that in turn reflect differences in DNA content, the difference in genome size between the helper and mutant should be as large as possible. A helper with wild-type genome length can be used for mutants with deletions greater than about 10% of the viral genome; helpers with genomes slightly larger than wild-type can be used for mutants with somewhat smaller deletions (*19*). It is important to ensure that no recombinants with a genome size nearer that of the mutant can arise during the growth of the mixed stock, because such recombinants make purification of the mutant more difficult.
3. The helper should have no properties that make low levels of contamination of the eventual purified mutant stock unacceptable, because no physical purification scheme is completely effective.

2. Materials

1. CsCl solutions: all CsCl solutions should be 20 mM Tris-HCl (final concentration), adjusted to pH 8.1. Adjustment of the pH should be made after dissolving the CsCl because some lots produce very acidic solutions:
   a. CsCl density 1.25: refractive index 1.3572; 33.8 g CsCl per 100 mL of solution.
   b. CsCl density 1.34: refractive index 1.3663; 46.0 g CsCl per 100 mL of solution.
   c. CsCl density 1.7: refractive index 1.3992; 95.1 g CsCl per 100 mL of solution.
2. 1,1,2-Trichlorotrifluoroethane (Sigma, cat. no. T5271).
3. 200X IGEPAL (Sigma, cat. no. I-3021): 10% IGEPAL CA-630 solution in water.
4. TE: 10 mM Tris-HCl, pH 8.1, 1 mM ethylene diamine tetraacetic acid (EDTA).
5. Phosphate-buffered saline (PBS) (per liter): 160 g NaCl, 4 g KCl, 18.2 g Na_2HPO_4, 4 g KH_2PO_4 37.2 g EDTA. pH should be 7.2.

3. Methods

3.1. Isolation of Defective Mutants by Complementation Plaquing

3.1.1. With Defective Helpers (see **Notes 1** and **2**)

1. Prepare host cell monolayers in 6-cm tissue culture dishes. If applicable, the host cells should be nonpermissive for the helper.
2. Determine the number of cells in one dish.

3. Transfect monolayers with mutant DNA by the calcium phosphate procedure (*see* Chapter 1). DNA fragments can be used if mutants are being constructed by a recombinational strategy.
4. After transfection, infect the monolayer at an MOI of 2–5 PFU/cell with the helper virus. Remove the medium from the transfected dishes, add the virus in 1 mL of medium, adsorb for 2 h, remove the inoculum, and fill the dishes with agar medium.
5. Continue by a standard plaquing protocol. If the helper is a *ts* mutant, incubate the dishes at the restrictive temperature.
6. When plaques are visible, pick and screen for the presence of the mutant (*see* Chapter 1).

3.1.2. With Nondefective Helpers

1. Prepare helper virus DNA–protein complex (DNAPC; *see* Chapter 1).
2. Mix mutant DNA and helper DNAPC in a molar ratio of 10:1 to 50:1 and transfect appropriate monolayers. The optimal amount of DNA for transfection varies depending on the plaque-forming efficiency of the helper DNA; adjust the DNA level to produce about 100 plaques per dish.
3. Treat as a normal plaque assay.
4. Pick and screen plaques for the presence of the mutant.

3.2. Growth of Mixed Stocks

1. Prepare host cell monolayers in 6-cm tissue culture dishes. If applicable, the host cells should be nonpermissive for the helper.
2. Remove the medium from each dish. Place the dishes with one edge slightly raised (for example, resting on a pencil) on a tray. This makes it possible to restrict the inoculum to a small area of the dish and raises the MOI in that region.
3. Pipet 0.1–0.25 mL of a mixed inoculum onto the lower edge of the monolayer. The inoculum can be a ministock (*see* Chapter 1), a portion of a previously made mixed stock, or a stock enriched for the deletion mutant by one round of CsCl density gradient centrifugation, diluted with medium to approx 10^9 infectious units per mL.
4. With the dish still tilted, incubate at 37°C in a humidified incubator for 2 h. (Proper humidification is important to prevent the raised side of the monolayer from drying out.)
5. Remove the inoculum, refill the dish with medium, place the dish flat, and incubate until all of the cells have detached from the plate. If a *ts* helper is being used, incubate at the restrictive temperature. Feed twice weekly by replacement of the medium until evidence of viral infection is seen over a substantial portion of the monolayer.
6. Harvest the infected cells and medium when all of the cells have detached from the dish. This mixed stock can be stored at –80°C until use.

3.3. Purification of Deletion Mutants From Mixed Stocks

Because adenovirus particles with differing DNA contents have differing buoyant densities in CsCl, adenovirus deletion mutants grown in the presence of helper virus can be separated from the helper by equilibrium sedimentation in CsCl density gradients. Only mutants with fairly large deletion mutations (>10%) can be efficiently purified from helpers with a wild-type genome size by this method, although helper virus with longer than wild-type genomes have been used to make the purification of mutants with smaller deletions possible *(19)*.

1. Prepare a mixed lysate. One dish should be labeled with ^{32}P, as described below.
2. To the mixed stock, add IGEPAL to a final concentration of 0.05%.
3. Extract the stock vigorously with 1/5 vol of 1,1,2-trichlorotrifluoroethane. Recover the aqueous phase after centrifugation at 4000g for 5 min in a Sorvall GSA rotor. Re-extract the cell debris and organic phase with a small volume of PBS; recover the aqueous phase and pool with the supernatant from the previous centrifugation.
4. Centrifuge the pooled supernatants at 10,000 rpm (g) for 5 min (Sorvall SS34 rotor) to remove small particulate cell debris.
5. Prepare a discontinuous CsCl gradient in a 35-mL polypropylene centrifuge tube by adding (in order) approx 20 mL extracted virus suspension, 4 mL of CsCl density 1.25, and 5 mL of CsCl density 1.7. Add each solution slowly through a pipet placed all of the way to the bottom of the tube. After the CsCl solutions have been added, fill the tube to the desired level with extracted virus suspension.
6. Centrifuge for 90 min at 29,000g (Sorvall SV288 rotor or equivalent) or 3 h at 82,000g (Beckman SW27 rotor or equivalent).
7. In a darkened tissue culture hood, illuminate the gradient with narrow beam of light from one side. A microscope lamp is a suitable light source. The virus will form a sharp, blue-white, translucent band at the interface of the two CsCl solutions. A broader, yellowish or tan, frequently granular layer of cell debris will appear at the top of the lighter CsCl cushion.
8. Collect the virus, avoiding the cell debris (*see* **Note 3**).
9. Adjust the concentrated virus suspension to a density of 1.34 (refractive index 1.3663) with 20 mM Tris-HCl, pH 7.5, or with the density 1.7 CsCl solution. Place in a centrifuge tube and fill to the required volume with CsCl density 1.34.
10. Centrifuge the suspension at 35,000 rpm (g) for 16 h in a Sorvall TV865 rotor (or equivalent). Two closely spaced virus bands should be visible in the center of the tube.
11. Fractionate the gradient into single-drop fractions through a hole made in the bottom of the tube.
12. Measure the radioactivity in each fraction by Cherenkov counting. Two more or less well-separated peaks should appear (**Fig. 1**).

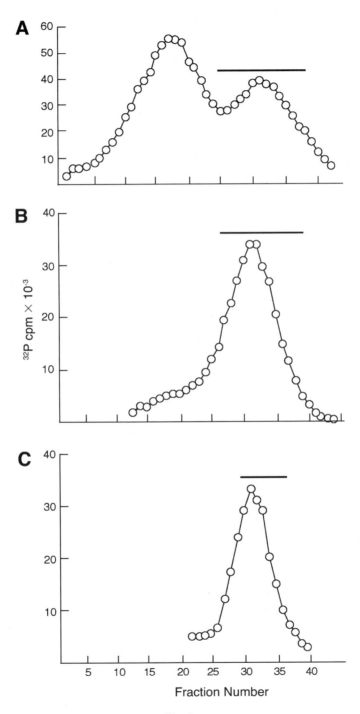

Fig 1.

13. Pool the fractions that comprise the lighter peak (*see* **Fig. 1**) and repeat **steps 7–10** (*see* **Note 4**).
14. Estimate the infectious titer of the virus suspension from its A_{260}. An A_{260} of 1 corresponds to a plaque-forming titer of 3.5×10^9 PFU/mL for Ad5 purified over three CsCl gradients.
15. The purity of the mutant stock can be assessed by plaquing under conditions permissive for the helper, or by restriction enzyme digestion of purified DNA.
16. Purified virus is stable for months in buoyant CsCl at 4°C. However, high-titer suspensions dialyzed against solutions of low ionic strength (e.g., TE) frequently precipitate. If it is necessary to remove the CsCl from a purified stock, first adjust the A_{260} of the suspension to 0.5 or lower and minimize storage time at low ionic strength.

3.4. Preparation of ³²P-Labeled Tracer Virus Particles (see Note 5)

1. Inoculate a 10-cm dish of cells with a mixed stock as described above.
2. Examine the dish daily, replacing the medium every 3 d until one-fourth to one-half of the cells show evidence of viral infection. Gently remove the medium from the dish and replace it with 10 mL of phosphate-free medium supplemented with 2% serum and 40 µCi/mL ³²P orthophosphate.
3. When all of the cells have become detached from the dish, harvest the cells and medium.
4. Collect the cells by low-speed centrifugation. Rinse the labeled cells twice by low-speed centrifugation and resuspension in PBS.
5. After the second rinse, resuspend the cells in 5 mL of PBS, add IGEPAL to 0.05%, and extract vigorously with 5 mL of 1,1,2-trichlorotrifluoroethane. Centrifuge and recover the aqueous phase. If intended for use as tracer, add this material to one tube of unlabeled virus and concentrate (**step 3**).

4. Notes

1. Some defective viruses kill cells that they infect even though they do not form plaques. If monolayers infected with helper at the MOI recommended here do not survive, the following modification should be used.
   a. Prepare monolayers in 24-well tissue culture dishes.
   b. Determine the number of cells in one well.

Fig 1. (*opposite page*) Purification of H2*dl*807 by CsCl density gradient centrifugation. A mixed lysate, lightly labeled with ³²P, containing H2*dl*807 and an Ad5 *ts* helper virus, was subjected to three successive bandings in CsCl density gradients. The radioactivity of single-drop fractions covering the middle portion of the gradients is shown. The top, center, and bottom panels represent the first, second, and third bandings, respectively. The mutant (upper) peaks have been aligned for clarity; the fraction numbers are arbitrary. The fractions pooled after each gradient are indicated by a black bar. H2*dl*807 lacks 12.5% of the viral genome. The contamination of the final H2*dl*807 pool with helper was approx 0.03%.

c. Transfect the wells with mutant DNA by the calcium phosphate procedure.

d. After transfection, infect the monolayer at an MOI of 2–5 with the helper virus. Remove the medium from the transfected dishes, add the virus in 0.5 mL of medium, adsorb for 2 h, remove the inoculum, and refill with medium. If the helper is a *ts* mutant, incubate at the restrictive temperature (*see* **Note 2**).

e. 12–16 h after infection, trypsinize each well and reseed the transfected/ infected cells in a 6-cm dish along with enough uninfected cells to form a confluent or nearly confluent monolayer.

f. After the cells have attached (8–24 h), overlay with agar medium and proceed as described in **Subheading 3.1.1., step 5**.

2. If the mutant is available as virus particles (as in the isolation of naturally occurring mutants), use one of the protocols in **Subheading 3.1.1.**, or in **Note 1**, replacing the transfection step with infection by the mutant stock at approx 50 infectious particles per dish (or well).

3. It is convenient to collect the virus through a hole made in the bottom of the tube with a pushpin. Plug the tube with a rubber stopper pierced by a large-gage syringe needle, close the needle with a finger over its hub, and make the hole in the bottom of the tube. The rate at which liquid flows out of the hole can be controlled by finger pressure on the hub of the syringe needle. Alternatively, a needle can be inserted through the side of the tube and the virus band drawn out with a syringe. For lysates from two or more 10-cm dishes, the virus band should be visible in the drops as they leave the tube; for small preps, ^{32}P-labeled virus tracer (*see* **Fig. 1**) can be added before centrifugation and fractions can be collected and the virus located by Cherenkov counting.

4. Depending on the purity required, two or more gradient steps may be necessary. In the experiment shown in **Fig. 1**, contamination of the deletion mutant stock by helper was about 0.03% after three gradient steps.

5. If labeled virus to be used as tracer is being prepared in parallel with a large unlabeled stock, the labeled cells should be harvested at the same time as the large stock, even if not all cells appear to be infected. If the labeled dishes are ready for harvesting before the remaining dishes, collect and rinse the cells and store at −80°C until needed.

References

1. Graham, F. L., Smiley, J., Russell, W. C., and Nairn, R. (1977) Characteristics of a human cell line transformed by DNA from human adenovirus type 5. *J. Gen. Virol.* **36,** 59–74.

2. Weinberg, D. H. and Ketner, G. (1983) A cell line that supports the growth of a defective early region 4 deletion mutant of human adenovirus type 2. *Proc. Natl. Acad. Sci. USA* **80,** 5383–5386.

3. Brough, D. E., Lizonova, A., Hsu, C., Kulesa, V. A., and Koveski, I. (1996) A gene transfer vector-cell line system for complete functional complementation of adenovirus early regions E1 and E4. *J. Virol.* **70,** 6497–6501.

4. Krougliak, V. and Graham, F. L. (1995) Development of cell lines capable of complementing E1, E4, and protein IX defective adenovirus type 5 mutants. *Hum. Gene Ther.* **6,** 1575–1586.

5. Wang, Q., Jia, X. C., and Finer, M. H. (1995) A packaging cell line for propagation of recombinant adenovirus vectors containing two lethal gene-region deletions. *Gene Ther.* **2,** 775–783.

6. Yeh, P., Dedieu, J. F., Orsini, C., Vigne, E., Denefle, P., and Perricaudet, M. (1996) Efficient dual transcomplementation of adenovirus E1 and E4 regions from a 293-derived cell line expressing a minimal E4 functional unit. *J. Virol.* **70,** 559–565.

7. Brough, D. E., Cleghon, V., and Klessig, D. F. (1992) Construction, characterization, and utilization of cell lines which inducibly express the adenovirus DNA-binding protein. *Virology* **190,** 624–634.

8. Amalfitano, A., Begy, C. R., and Chamberlain, J. S. (1996) Improved adenovirus packaging cell lines to support the growth of replication-defective gene-delivery vectors. *Proc. Natl. Acad. Sci. USA* **93,** 3352–3356.

9. Schaack, J., Guo, X., Ho, W. Y., Karlok, M., Chen, C., and Ornelles, D. (1995) Adenovirus type 5 precursor terminal protein-expressing 293 and HeLa cell lines. *J. Virol.* **69,** 4079–4085.

10. Oualikene, W., Lamoureux, L., Weber, J. M., and Massie, B. (2000) Protease-deleted adenovirus vectors and complementing cell lines: potential applications of single-round replication mutants for vaccination and gene therapy. *Hum. Gene Ther.* **11,** 1341–1353.

11. Zhang, W. and Imperiale, M. J. (2003) Requirement of the adenovirus IVa2 protein for virus assembly. *J. Virol.* **77,** 3586–3594.

12. Challberg, S. S. and Ketner, G. (1981) Deletion mutants of adenovirus 2: isolation and initial characterization of virus carrying mutations near the right end of the viral genome. *Virology* **114,** 196–209.

13. Kochanek, S. (1999) High-capacity adenoviral vectors for gene transfer and somatic gene therapy. *Hum. Gene Ther.* **10,** 2451–2459.

14. Cao, H., Koehler, D. R., and Hu, J. (2004) Adenoviral vectors for gene replacement therapy. *Viral Immunol.* **17,** 327–333.

15. Pastore, L., Morral, N., Zhou, H., Garcia, R., Parks, R. J., Kochanek, S., Graham, F. L., Lee, B., and Beaudet, A. L. (1999) Use of a liver-specific promoter reduces immune response to the transgene in adenoviral vectors. *Hum. Gene Ther.* **10,** 1773–1781.

16. Kochanek, S., Clemens, P. R., Mitani, K., Chen, H. H., Chan, S., and Caskey, C. T. (1996) A new adenoviral vector: Replacement of all viral coding sequences with 28 kb of DNA independently expressing both full-length dystrophin and beta-galactosidase. *Proc. Natl. Acad. Sci. USA* **93,** 5731–5736.

17. Fisher, K. J., Choi, H., Burda, J., Chen, S. J., and Wilson, J. M. (1996) Recombinant adenovirus deleted of all viral genes for gene therapy of cystic fibrosis. *Virology* **217,** 11–22.

18. Mitani, K., Graham, F. L., Caskey, C. T., and Kochanek, S. (1995) Rescue, propagation, and partial purification of a helper virus-dependent adenovirus vector. *Proc. Natl. Acad. Sci. USA* **92,** 3854–3858.

19. Falgout, B. and Ketner, G. (1987) Adenovirus early region 4 is required for efficient virus particle assembly. *J. Virol.* **61,** 3759–3768.

3

Construction of Adenovirus Type 5 Early Region 1 and 4 Virus Mutants

Peter Groitl and Thomas Dobner

Summary

This chapter describes a novel strategy that simplifies the generation and production of adenovirus type 5 (Ad5) mutants carrying defined mutations in early transcription units 1 (E1) and 4 (E4). The strategy involves three recombinant plasmids containing E1 (pE1-1235), E4 (pE4-1155), or the wild-type genome that lacks a portion of E3 (pH5pg4100). To generate recombinant viruses, mutations are first introduced into pE1- and/or pE4-transfer plasmids by site-directed mutagenesis. The mutagenized constructs are then ligated into plasmid pH5pg4100 containing the Ad backbone by direct cloning. Infectious viral DNAs are released from the recombinant plasmids by *Pac*I-digestion and transfected into the complementing cell lines 293 or W162, and viral progeny are isolated and amplified. The advantages of this strategy are multiple: all cloning steps are carried out in *Escherichia coli*, and any genetic region of the viral E1 and/or E4 transcription units can be specifically modified or deleted. Moreover, foreign genes can be introduced into the E1 and/or E4 regions, and expression of viral or therapeutic genes can be controlled by cell-type specific and/or inducible promoters.

Key Words: Adenovirus type 5; early region 1; E1; early region 4; E4; E1B-55K; E4orf6.

1. Introduction

The adenovirus type 5 (Ad5) E1B-55K and E4orf6 proteins are multifunctional regulators of Ad5 replication participating in many processes required for maximal virus production. A complex containing the two proteins has been implicated in the proteolytic degradation of the tumor suppressor protein p53 and the Mre11/Rad50/Nbs1 (MRN) complex, shut-off of host-cell protein synthesis, as well as selective viral late mRNA transport from the nucleus to the cytoplasm (reviewed in **refs.** *1,2*). A certain amount of recent data suggests that most of these activities may involve a CRM1-dependent nuclear export pathway because both proteins continuously shuttle between the nuclear and

From: *Methods in Molecular Medicine, Vol. 130:*
Adenovirus Methods and Protocols, Second Edition, vol. 1:
Adenoviruses Ad Vectors, Quantitation, and Animal Models
Edited by: W. S. M. Wold and A. E. Tollefson © Humana Press Inc., Totowa, NJ

cytoplasmic compartments via leucine-rich nuclear export signals (NESs) *(3,4)* and because mutations in the E4orf6 NES negatively affect late viral gene expression in artificial transfection/infection complementation assays *(5)*. In contrast, a different conclusion was drawn from similar studies showing that E1B-55kDa/E4orf6 promote late gene expression without intact NESs or active CRM1 *(6)*. To evaluate the role of CRM1-dependent nuclear export of E1B-55kDa/ E4orf6 for efficient late viral gene expression and virus production in the normal context of Ad5-infected cells, we have developed a novel method allowing the construction of single and double virus mutants carrying defined amino acid substitutions in the NESs of E1B-55K and/or E4orf6 proteins by direct cloning of genetically modified E1 and/or E4 cassettes into an Ad5 backbone.

Figure 1 outlines the strategy of our approach for the construction of these viruses. The strategy involves three recombinant plasmids containing Ad5 E1 from nt 1-5766 (pE1-1235) or E4 from nt 32840-35934 (pE4-1155) and the Ad5 wild-type genome that lacks a portion of E3 (pH5*pg*4100). These plasmids were constructed through several rounds of subcloning and modification of restriction endonuclease fragments. Plasmids pE1-1235 and pE4-1155 are used to introduce mutations into the E1 and E4 genes by site-directed mutagenesis, respectively, while plasmid pH5*pg*4100 serves as the DNA backbone to recombine the mutations into the Ad5 chromosome by direct cloning. For the generation of E1 and/or E4 mutant viruses and their wild-type parent H5*pg*4100, infectious viral DNAs are released from the recombinant plasmids by *Pac*I digestion and transfected into the complementing cell lines 293 *(7)* or W162 *(8)*. Viral progeny are isolated and amplified in 293 or W162 cells. Sub-

---

Fig. 1. (*opposite page*) Constructions of E1 and/or E4 virus mutants. (**A**) Schematic diagram of adenoviral plasmid pH5*pg*4100. The Ad genome in pH5*pg*4100 is identical to the Ad5 wild-type sequence (GenBank accession no. AY339865) except that it lacks a portion of E3 (nt 28593-30471), and contains a unique *Bst*BI endonuclease restriction site at nt 30955 preceding E4. The left and right inverted terminal repeats (*l*ITR and *r*ITR) of the viral DNA are flanked by a *Pac*I site, allowing the excision of the viral DNA (34054 bp) from the pPG-S2 vector by *Pac*I digestion. Endonuclease restriction sites in pPG-S2 used for construction of recombinant Ads are underlined. ΔE3 indicates the 1879 bp deletion in E3. (**B**) Schematic outline of cloning strategy employed to generate recombinant Ads. Mutations are introduced into the E1 or E4 genes in pE1-1235 or pE-1155, respectively, by site-directed mutagenesis. These plasmids and pH5*pg*4100 are linearized with either *Swa*I plus *Bst*Z17I (pE1-1235) or *Bst*BI (pE-1155), and the corresponding fragments from pH5*pg*4100 are replaced with the appropriate fragments from mutagenized pE1 or E4 constructs. Viral genomes are released by *Pac*I digestion and used for transfection of the complementing cell lines 293 or W162. Viruses are then isolated and propagated by using 293 or W162 cells.

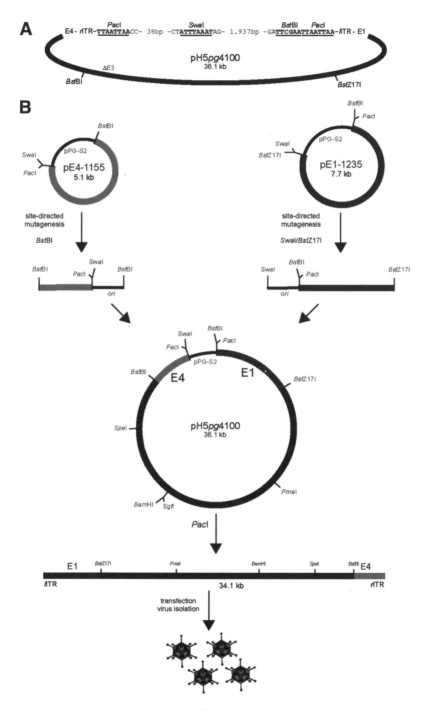

Fig. 1.

sequently, the integrity of the recombinant viruses is verified by restriction digestion and DNA sequencing.

2. Materials

1. 10X Tris borate EDTA electrophoresis buffer: 108 g Tris base, 55 g boric acid, 4 mL 0.5 M ethylene diamine tetraacetic acid (EDTA), pH 8.0, per liter. Store at room temperature.
2. 0.6% Agarose in 1X TBE for gel electrophoresis.
3. Ethidium bromide stock solution (10 mg/mL): store at 4°C.
4. 1 M Tris-HCl , pH 8.0: autoclave and store at room temperature.
5. Herring sperm DNA: dissolve herring sperm DNA in 10 mM Tris-HCl, pH 7.0, and sonicate until the viscosity of the solution no longer changes appreciably. Measure the concentration and adjust it to 1 mg/mL with 10 mM Tris-HCl, pH 7.0. Store at –20°C.
6. 2.5 M CaCl$_2$: filter-sterilize and store at –20°C.
7. 2X Balanced electrolyte solution (BES)-buffered saline: dissolve 1.07 g of BES (*N,N-bis* [2-hydroxymethyl]-2-aminoethanesulfonic acid), 1.6 g NaCl, and 0.027 g of Na$_2$HPO$_4$ in a total volume of 90 mL of distilled H$_2$O. Adjust the pH of the solution to 6.96 with HCl at room temperature, and adjust the volume to 100 mL with distilled H$_2$O. Filter-sterilize and store in aliquots at –20°C.
8. 20X Phosphate-buffered saline (PBS): dissolve 4 g KCl, 4 g KH$_2$PO$_4$, 160 g NaCl, 22.8 g Na$_2$HPO$_4$ to a final volume of 1 L in water. Autoclave and store at room temperature.
9. 100 mM HEPES (*N*-2-hydroxymethylpiperazine-*N*'-2-ethanesulfonic acid) (100 mL): dissolve 2.383 g in H$_2$O and adjust pH with NaOH to 7.4. Store at 4°C and filter-sterilize aliquots before use.
10. 2X HEPES-buffered saline (modified): dissolve 1.6 g of NaCl, 0.074 g of KCl, and 1.19 g of HEPES in a total volume of 90 mL distilled H$_2$O. Adjust the pH with NaOH to 7.4 and the volume to 100 mL with distilled H$_2$O. Store at 4°C.
11. 1.25 g/cm^3 CsCl solution: dissolve 36.16 g CsCl in 100 mL HEPES-buffered saline (modified). Filter-sterilize and store at room temperature.
12. 1.4 g/cm^3 CsCl solution: dissolve 62 g CsCl in 100 mL HEPES-buffered saline (modified). Filter-sterilize and store at room temperature.
13. 5 M NaCl (100 mL): dissolve 29.22 g NaCl in distilled H$_2$O. Autoclave and store at room temperature.
14. Bovine serum albumin (BSA) stock solution: dissolve 1 g of BSA in 100 mL distilled H$_2$O. Filter-sterilize and store aliquots at –20°C.
15. Virus storage buffer (100 mL): 12 mL of 100 mM HEPES stock solution (pH 7.4), 2.4 mL of 5 M NaCl stock solution, and 50 mL glycerol. Add 1 mL BSA from a 10 mg/mL stock solution and adjust volume to 100 mL with distilled water. Filter-sterilize and store aliquots at 4°C.
16. Proteinase K (10 mg/mL): resuspend 100 mg in 10 mL of 10 mM Tris-HCl, pH 8.0, and digest for 20 min at 37°C. Store aliquots at –20°C.
17. 10% Sodium dodecyl sulfate (SDS): dissolve 10 g to a total volume of 100 mL in sterile H$_2$O. Do not autoclave. Store at room temperature.

18. Plasmids (*see also* **Fig. 1**):
    a. pH5*pg*4100: Bacmid containing Ad5 genome (36,067 bp) inserted into the *Pac*I site of the bacterial cloning vector pPolyII *(9)*. The viral genome lacks nucleotides (nt) 28,593–30,471 (encompassing E3) and contains two additional unique endonuclease restrictions sites at nt 5764 (*Bst*Z17I) and nt 30955 (*Bst*BI) (throughout this chapter nucleotide numbering is according to the published Ad5 sequence from GenBank, accession no. AY339865).
    b. pE1-1235: E1-transfer vector (7734 bp) containing Ad5 E1 region from nt 1-5766 in vector pPG-S3.
    c. pE4-1155: E4-transfer vector (5090 bp) containing Ad5 E4 region nt 32,840–35,934 in vector pPG-S2.
    d. The complete sequences of pE1-1235, pE4-1155 and pH5*pg*4100 (36067 bp) were determined on both strands by sequence-derived oligonucleotide primers and is available upon request.
19. In vitro site-directed mutagenesis kit (Stratagene).
20. Ultracompetent XL2-BLUE cells from (Stratagene; cat. no. 200150).
21. LB-amp plates (100 µg ampicillin/mL).
22. Cell lines 293 and/or W162.
23. Dulbecco's modified Eagle's medium (DMEM) supplemented with 10% fetal calf serum (FCS), 100 U of penicillin, and 100 µg of streptomycin per mL.
24. 60-mm Falcon tissue culture dishes (cat. no. 353004), 100-mm Falcon tissue culture dishes (cat. no. 353003), 150-mm Falcon tissue culture dishes (cat. no. 353025).
25. 5-mL Falcon polystyrene round-bottom tubes (cat. no. 352052).
26. 50-mL Polypropylene tubes.
27. PBS.
28. Thin-walled 14-mL polyacetylene tubes (Herolab; cat. no. 252360).
29. Nunc CryoTube™ Vials (cat. no. 343958).
30. RNase T_1 (Roche; cat. no. 109193).

3. Methods

3.1. Construction of Ad5 Recombinants

1. To generate Ad5 mutants carrying defined mutations in the E1B and E4 genes, point mutations are first introduced into pE1-1235 and pE4-1155 by site-directed mutagenesis (In Vitro Site-Directed Mutagenesis Kit, STRATAGENE) with oligonucleotide primers.
2. Verify mutations from purified plasmids by DNA-sequencing.
3. Digest the mutagenized constructs with the appropriate restriction enzyme(s): pE1-1235 with *Swa*I/*Bst*Z17I and pE4-1155 with *Bst*BI.
4. Digest pH5*pg*4100 bacmid DNA with the same restriction enzyme(s).
5. Purify DNA fragment (*see* **Note 1**):
    a. Load 20 µg of digested pH5*pg*4100 bacmid DNA on a preparative 0.6% agarose gel containing 0.5 µg ethidium bromide/1 mL gel solution (not more than 1–2 µg DNA/1 cm running front of the gel).

    b. Run the gel overnight at 40 V to avoid mechanical stress on the DNA.

    c. Cut out the upper DNA fragment under weak UV (365 nm) using a scalpel and transfer it to 1.5-mL Eppendorf tubes.

    d. Spin the tubes in a SORVALL RC 5C Plus centrifuge (SS-34 rotor) at 50,200g for 90 min at 4°C using Eppendorf tube adaptors.

    e. Immediately transfer the supernatants to fresh Eppendorf tubes.

    f. Precipitate the DNA with isopropanol.

    g. Wash twice with 70% ethanol.

    h. Air-dry the pellets briefly and resuspend the DNA in 40 µL 10 mM Tris-HCl, pH 8.0.

    i. Check an aliquot on agarose gel.

    j. The compressed agarose after **step e** still contains a lot of DNA and might be allowed to swell again in sterile ddH$_2$O overnight. Afterwards repeat the procedure from **step d**.

6. Dephosphorylate the linearized pE1-1235 or pE4-1155 plasmids and check integrity of the DNA with an aliquot on an agarose gel.

7. Ligate 10–20 ng of pE1-1235 or pE4-1155 with a threefold excess of ends from the linearized pH5*pg*4100 backbone at 13°C overnight.

8. Transform ultracompetent XL2-BLUE cells (Stratagene) exactly as described by the manufacturer and spread on LB-amp plates (100 µg ampicillin/mL) (*see* **Note 2**).

9. Incubate the plates at 37°C for at least 24 h to be able to determine the size of the colonies (*see* **Note 3**).

10. Inoculate small, well-isolated colonies overnight each in 5 mL LB-amp medium (100 µg ampicillin/mL) at 37°C in a shaker (220 rpm).

11. Perform minipreps from 1 mL of each culture. Store the remainder at 4°C.

12. Digest one-half of the DNAs with *Pac*I and separate the fragments on a 0.6% agarose gel. Check the orientation of the insert.

13. Digest the other half of the DNAs from the positive clones with *Hin*dIII and separate the fragments on a large 0.6% agarose gel (*see* **Fig. 2**). Check the resulting restriction patterns for deletions or transposon integration.

14. Spin down the remainder of the culture from the positive clones to make glycerol stocks.

3.2. Transfection of Viral DNA

1. The cell lines 293 and W162 are grown as monolayers in DMEM supplemented with 10% FCS, 100 U of penicillin, and 100 µg of streptomycin per mL in a 7% CO$_2$ atmosphere at 37°C (*see* **Note 4**).

2. For transfection of viral DNA, grow cells to 50–70% confluency in 60-mm tissue culture dishes. Use cells at a low passage number.

3. Digest the recombinant bacmid DNA with *Pac*I to remove the bacterial vector (pPG-S2) from the Ad5 genome.

4. Precipitate the DNA with isopropanol and wash with 70% ethanol.

5. Air-dry the DNA briefly and resuspend in 10 mM Tris-HCl, pH 7.0, to a final concentration of 1 µg/µL.

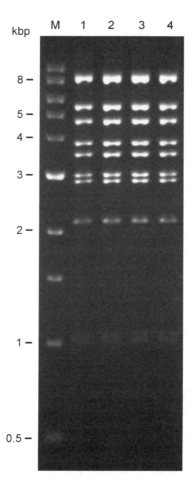

Fig. 2. Restriction pattern of wild-type H5*pg*4100 and mutant virus DNA. Viral DNA was isolated from purified virus particles as described (*see* **Subheading 3.5.**). A 0.8-µg sample of wild-type (lane 1) and mutant virus DNA (lanes 2–4) was digested with *Hin*dIII and separated on a 0.6% agarose gel. M, marker.

6. Mix 5 µg of viral and 15 µg of sonicated herring sperm DNA with sterile ddH$_2$O and 2.5 *M* CaCl$_2$ to a final volume of 0.5 mL and a concentration of 0.25 *M* CaCl$_2$.
7. Per transfection, aliquot 0.5 mL 2X BES-buffered saline in 5-mL polystyrene round-bottom tubes (Falcon). Add dropwise the DNA solution under constant but slow vortexing (*see* **Note 5**).
8. Incubate the mixture for 5 min at room temperature.
9. Remove the medium from the dishes of cells and wash once with prewarmed PBS.
10. Give prewarmed fresh medium to the dishes of cells and add, dropwise, the CaCl$_2$/DNA/BES-buffered saline solution. Gently agitate the dishes to distribute the DNA evenly.

11. Incubate overnight at 37°C in a humidified incubator in an atmosphere of 5–7% CO_2.
12. Change the medium and incubate the dishes of cells for a further 4–7 d.

3.3. Propagation of Ad5 Virus Mutants

1. Harvest the transfected cells using a rubber policeman or cell scraper, transfer them to 15-mL polypropylene test tubes, and spin them down (2000 rpm [800g]; 5 min at 4°C).
2. Remove the supernatant and resuspend the cells in 6 mL of DMEM lacking both antibiotics and FCS.
3. Freeze in liquid nitrogen for a few minutes and then thaw in a 37°C water bath. Repeat these freeze–thaw steps twice. The frozen cells can be kept at –80°C for several months until continuing with the procedure.
4. Centrifuge the test tubes for 10 min at 3000g and room temperature.
5. Take 1 mL of the supernatant for a control infection and the remaining 5 mL rest for virus propagation.
6. Infect 50% confluent monolayers of the appropriate complementing cell line in 100-mm-diameter tissue culture dishes with 5 mL of the virus-containing supernatant.
7. Incubate for 3 h at 37°C in a humidified incubator in an atmosphere of 5–7% CO_2 under constant but gentle agitation on a shaker. Alternatively agitate the dishes every 5–10 min by hand.
8. Remove the infection medium and add DMEM supplemented with antibiotics and 10% FCS.
9. Incubate for 3–5 d. Check every day under the microscope to determine whether a cytopathic effect has become visible. As soon as the majority of the cells are rounding up and are losing contact to the surface of the tissue culture dishes, the cells are ready for harvesting. This might be the case in the first round after transfection or in a later round of infection. If no clear cytopathic effect is visible, harvest the cells after 5 d and repeat **step 1** from **Subheading 3.3.**
10. When there is a cytopathic effect affecting the majority of the cells, harvest the cells, freeze–thaw cells three times in 9 mL of DMEM, and infect 50% confluent monolayers in 150-mm-diameter tissue culture dishes. Continue infection as in **Subheading 3.3., steps 7–9.**
11. After having harvested a completely infected monolayer from a 150-mm-diameter dish, resuspend the cells in 10 mL DMEM, freeze–thaw cells three times in 50-mL polypropylene tubes, and infect 10 50–80% confluent monolayers in 150-mm-diameter dishes as in **Subheading 3.3., step 10**.

3.4. Large-Scale Ad Purification

1. Harvest infected cells from 10 150-mm culture dishes, transfer cells to 50-mL polypropylene centrifuge tubes, and spin them down (2000 rpm [800g]; 5 min at 4°C).
2. Pool the infected cells in one centrifuge tube and wash gently in PBS.
3. Pellet the infected cells again, aspirate the supernatant, and resuspend in 7 mL of 100 mM HEPES, pH 7.4.

4. Freeze the infected cells in liquid nitrogen for a few minutes and thaw cells in a 37°C water bath. Repeat these freeze–thaw steps twice.
5. Pellet the debris at 3000–5000*g* for 10 min at 15°C.
6. Prepare a discontinuous CsCl density gradient in a thin-walled 14-mL polyacetylene tube. Aliquot 2 mL of CsCl solution (1.4 g/cm^3 in HEPES-buffered saline) into each tube and mark the surface on the wall of the tube. Pipet slowly 2.5 mL CsCl solution (1.25 g/cm^3 in HEPES-buffered saline) on top of the first solution.
7. Carefully load supernatant from **Subheading 3.4., step 5**, on top of the CsCl density gradient.
8. Centrifuge for 90 min at 130,000*g* and 15°C in a swing-out SW40 rotor.
9. After centrifugation, a bluish-white virus band should be visible a little above the marking on the tube.
10. Collect the viral band by side puncture with a syringe or by pipetting from the top (first remove the waste on top of the gradient with a 10 mL pipet, then collect the virus band using a blue tip).
11. For a concentrated virus stock, measure the volume by pipetting and dilute six-fold by adding 5 vol of 1.2X virus storage buffer.
12. Aliquot the virus stock into several 1.8-mL cryovials. At –80°C the frozen stock will be stable for years; the liquid working stock can be kept for several months at –20°C.

3.5. Characterization of Viral DNA

1. To verify the presence of mutations and to confirm the integrity of the viral DNA by DNA sequencing and restriction endonuclease digestion, dialyze a 0.5- to 1-mL aliquot of the purified virus stock for several hours (or overnight) against a large volume of PBS or Tris-EDTA (TE) buffer. Change buffer twice.
2. Transfer the virus solution into a 1.5-mL test tube and add Proteinase K to a final concentration of 200 μg/mL and SDS to a final concentration of 0.5%. Add 500–1000 U RNase T$_1$ (Roche; cat. no. 109193).
3. Incubate for at least 3 h or overnight at 37°C.
4. Perform a PCIA extraction (phenol:chloroform:isoamyl alcohol = 25:24:1), and precipitate the DNA afterwards.
5. Resuspend the DNA in 10 m*M* Tris-HCl, pH 8.0, and measure the concentration of the DNA.
6. Digest 0.8 μg DNA with *Hin*dIII, and separate the fragments on a large 0.6% agarose gel with 0.5 μg ethidium bromide/mL gel solution. Check the resulting characteristic restriction pattern for deletions or other rearrangements.
7. Use 1–2 μg virus DNA and 15–25 p*M* primer per virus DNA sequence reaction.

4. Notes

1. Restriction of pH5*pg*4100 with *Swa*I/*Bst*Z17I or *Bst*BI yields a large DNA fragment of approx 28 or 31 kb, respectively, and a smaller fragment of either 7.7 kb (*Swa*I/*Bst*Z17I) or 5.1 kb (*Bst*BI). In particular, the quality of the large DNA fragment containing the Ad5 backbone is most critical for the successful cloning

of the Ad5 recombinant DNA. Thus, use general precautions to prevent nicks in the DNA fragment, which may result, for example, from vigorous vortexing. In order to keep UV damage of the DNA to a minimum, we usually add guanine-9-β-D-ribofuranoside (Fluka; cat. no. 51050) to a final concentration of 1 mM to both the agarose gel solution and the TBE running buffer *(10)*.

2. For transformation of the recombinant Ad plasmids, we strongly recommend the use of commercially available highly competent cells, such as XL2-Blue from Stratagene. In our experience this prokaryotic host strain reproducibly allows transformation of the large Ad DNA-containing plasmids (~36 kb) with high efficiency. In addition, XL2-Blue is recombination-deficient, which substantially reduces mutations (deletions and transposon integration) in the Ad DNA during the following amplification process.

3. The transformants on the LB-amp plates appear as large and tiny colonies. Do not pick the large-sized colonies because they mostly contain truncated forms of the recombinant Ad plasmids.

4. The cell line W162 was generated by introducing the whole Ad5 E4 region, including its natural promoter into monkey Vero cells *(8)*. W162 cells supply all E4 functions *in trans* and are used for transfections of viral DNA and propagation of Ads containing mutations in E4. 293 cells are derived from human embryonic kidney cells transformed by sheared Ad5 DNA *(7)*. They produce the Ad5 E1A and E1B genes under the control of their endogenous promoters *in trans* and are used for transfections of viral DNA and propagation of Ads containing mutations in E1A and/or E1B, allowing the production of infectious virus particles when cells are transfected with Ad DNA carrying mutations in E1A and/or E1B. In order to achieve optimal transfection efficiencies and titers, it is important that both cell lines are healthy and plated at optimal density. In particular, 293 cells should be used at low passage number (passage 20–30), and passaged at 40% confluency. At late passages 293 cells exhibit a shorter life span, do not adhere well to the tissue culture dishes, and do not transfect efficiently.

5. Alternatively, air can be bubbled into the 2X BES-buffered saline while the DNA solution is added. Other transfection methods using liposome micelles (e.g., Lipofectamine 2000, Invitrogen) may also be used, but the calcium phosphate method has been most consistently successful in our experience.

References

1. Täuber, B. and Dobner, T. (2001) Molecular regulation and biological function of adenovirus early genes: the E4 ORFs. *Gene* **278,** 1–23.
2. Flint, S. J. and Gonzalez, R. A. (2003) Regulation of mRNA production by the adenoviral E1B 55-kDa and E4 Orf6 proteins. *Curr. Top. Microbiol. Immunol.* **272,** 287–330.
3. Dobbelstein, M., Roth, J., Kimberly, W. T., Levine, A. J., and Shenk, T. (1997) Nuclear export of the E1B 55-kDa and E4 34-kDa adenoviral oncoproteins mediated by a rev-like signal sequence. *EMBO J.* **16,** 4276–4284.

4. Krätzer, F., Rosorius, O., Heger, P., et al. (2000) The adenovirus type 5 E1B-55k oncoprotein is a highly active shuttle protein and shuttling is independent of E4orf6, p53 and Mdm2. *Oncogene* **19,** 850–857.
5. Weigel, S. and Dobbelstein, M. (2000) The nuclear export signal within the E4orf6 protein of adenovirus type 5 supports virus replication and cytoplasmic accumulation of viral mRNA. *J. Virol.* **74,** 764–772.
6. Rabino, C., Aspegren, A., Corbin-Lickfett, K., and Bridge, E. (2000) Adenovirus late gene expression does not require a rev-like nuclear RNA export pathway. *J. Virol.* **74,** 6684–6688.
7. Graham, F. L., Smiley, J., Russell, W. C., and Nairn, R. (1977) Characteristics of a human cell line transformed by DNA from human adenovirus type 5. *J. Gen. Virol.* **36,** 59–72.
8. Weinberg, D. H. and Ketner, G. (1983) A cell line that supports the growth of a defective early region 4 deletion mutant of human adenovirus type 2. *Proc. Natl. Acad. Sci. USA* **80,** 5383–5386.
9. Chartier, C., Degryse, E., Gantzer, M., Dieterle, A., Pavirani, A., and Mehtali, M. (1996) Efficient generation of recombinant adenovirus vectors by homologous recombination in Escherichia coli. *J. Virol.* **70,** 4805–4810.
10. Gründemann, D., and Schömig, E. (1996) Protection of DNA during agarose gel electrophoresis against damage induced by ultraviolet light. *BioTechniques* **21,** 898–903.

4

Construction of Mouse Adenovirus Type 1 Mutants

Angela N. Cauthen, Amanda R. Welton, and Katherine R. Spindler

Summary

Mouse adenovirus provides a model for studying adenovirus pathogenesis in the natural host. The ability to make viral mutants allows the investigation of specific mouse adenoviral gene contributions to virus–host interactions. Methods for propagation and titration of wild-type mouse adenovirus, production of viral DNA and viral DNA–protein complex, and transfection of mouse cells to obtain mouse adenovirus mutants are described in this chapter. Plaque purification, propagation, and titration of the mutant viruses are also presented.

Key Words: Mouse adenovirus type 1; virus purification; transfection; plaque assay; DNA–protein complex; virus mutants.

1. Introduction

Construction of mouse adenovirus type 1 (MAV-1) mutants has facilitated studies of adenoviral pathogenesis in the natural host. We have isolated viral mutants of MAV-1 early regions 1A (E1A) and 3 (E3) *(1–4)*. These mutants have altered lethality in adult mice, as measured by 50% lethal dose (LD$_{50}$) assays. E1A mutants have increased sensitivity to type I and type II interferon in vitro relative to wild-type (wt) virus *(5)*. MAV-1 E1A interacts with mouse Sur2, a subunit of mediator complex, and E1A viral mutants were used to confirm that the interaction maps to E1A conserved region 3 (CR3) *(6)*. E3 mutant infections result in different histopathology from wt MAV-1, exhibiting less inflammation in the brain in inbred and outbred mice than wt virus (L. Gralinski and K. Spindler, unpublished; *see also* **ref. 2**).

Wild-type MAV-1 causes acute and persistent infections in mice (reviewed in **ref. 7**). Wild-type and mutant MAV-1 can be propagated in cell lines and primary mouse cells in vitro. MAV-1 is the better characterized of the two known mouse adenovirus serotypes. It causes a fatal disease in newborns or

From: *Methods in Molecular Medicine, Vol. 130:*
Adenovirus Methods and Protocols, Second Edition, vol. 1:
Adenoviruses Ad Vectors, Quantitation, and Animal Models
Edited by: W. S. M. Wold and A. E. Tollefson © Humana Press Inc., Totowa, NJ

adult nude or severe combined immune-deficient mice when inoculated intra-peritoneally, intracerebrally, or intranasally *(8–10)*. Infection of some strains of adult mice with high dosages of MAV-1 results in death for all of the animals *(11–13)*. Virus is disseminated throughout many organs and found at the highest levels in the spleen and brain *(12,14)*. We and others have documented an acute central nervous system disease in outbred and inbred mice in which brains and spinal cords exhibit encephalomyelitis *(11,12)*.

MAV-1 has a genome organization like that of the human adenoviruses: a set of genes is transcribed early after infection, and a set of late genes is transcribed after the onset of viral DNA replication (reviewed in **ref.** *15*). The DNA genome of MAV-1 consists of 30,944 base pairs, and it has been completely sequenced *(16)*. Like the human adenoviruses, a terminal protein is covalently attached to the 5'-end of each strand *(17,18)*.

The method described here for constructing mouse adenovirus mutants relies on the ability of MAV-1 to undergo homologous recombination in mouse cells and is largely based on methods used to make human adenovirus mutants *(19–21)*. Although there is a full-length infectious plasmid of MAV-1 that could be used to generate mutants *(22)*, the low efficiency of wt virus production from it has limited its utility. Thus, we describe here the method that has been successfully used to make all MAV-1 mutants to date. In the first section we describe the preparation of MAV-1 DNA–protein complex. Preparation of stocks of MAV-1 is slightly different because of the fact that MAV-1 is released into the medium and does not remain associated with cell debris *(23)*. Thus, the virus stock is first concentrated using polyethylene glycol *(24)* and then purified in CsCl gradients *(25)*. DNA–protein complex is prepared from the virus *(26)* and then digested with *Eco*RI. To increase the probability of obtaining recombinant viruses relative to wt viruses, we added a restriction digest/Klenow fill-in step to the preparation of DNA–protein complex prior to transfection *(1,4)*. We also describe the preparation of MAV-1 viral DNA (*see* **Note 6**).

In the second section we describe the transfection protocol. We have successfully used mouse 3T6 cells to obtain mutants. In some cases we have also used 3T6 cell derivatives, which inducibly express the gene region to be mutated *(3,4)*. We use a calcium phosphate transfection protocol modified from that of Gorman et al. *(27)*, but other methods may be used. Any of a variety of methods can be used to obtain the mutation in a plasmid containing the gene of interest, including oligonucleotide-directed mutagenesis, polymerase chain reaction (PCR) mutagenesis, and restriction fragment replacement cloning.

In the third section we describe the identification and plaque purification of mutants. We have used preparation of viral DNA directly from plaques, from cell culture fluid, and by the Hirt method *(28)* for use in PCR. We have also used DNA prepared by the Hirt method for direct restriction enzyme analysis of potential mutants. In the final sections we describe the preparation of MAV-1 viral stocks and our plaque assay protocol.

2. Materials

1. 5 *M* NaCl.
2. 50% PEG (PEG8000, Sigma P-2139): weigh out 250 g of PEG and add water to 500 mL. Stir until dissolved and autoclave. Mix while cooling to prevent precipitation of the PEG. Store at room temperature.
3. 1 *M* Tris-HCl, pH 8.0. Autoclave and store at room temperature.
4. 13-mL Ultracentrifuge tubes (Seton 7030).
5. 1 *M* Tris-HCl, pH 7.4. Autoclave and store at room temperature.
6. CsCl $\rho = 1.2$ g/cm^3 = 22.49%: 26.99 g/100 mL of 50 m*M* Tris-HCl, pH 7.4. Store at room temperature.
7. CsCl $\rho = 1.4$ g/cm^3 = 38.60%: 54.04 g/100 mL of 50 m*M* Tris-HCl, pH 7.4. Store at room temperature.
8. Gradient dripper.
9. Peristaltic pump.
10. Gradient maker (10-mL capacity for each chamber).
11. 12,000–14,000 Molecular-weight exclusion dialysis tubing. Store in 10 m*M* EDTA at 4°C.
12. 15-mL Conical polypropylene tubes.
13. 0.5 *M* EDTA, pH 8.0: autoclave and store at room temperature.
14. 8 *M* Guanidine HCl. Store at room temperature.
15. Optical grade CsCl (Pharmacia 17-0845-02).
16. 1.5-mL Microfuge tubes.
17. Tris-EDTA (TE): 10 m*M* Tris-HCl, pH 8.0, 1 m*M* ethylene diamine tetraacetic acid (EDTA), pH 8.0. Store at room temperature.
18. *Eco*RI (20 U/µL).
19. 10X *Eco*RI buffer: 1 *M* NaCl, 0.5 *M* Tris-HCl, pH 7.5, 0.1 *M* MgCl$_2$, 10 m*M* dithiothreitol (DTT). Add DTT before using. Store at room temperature.
20. 100 m*M* DTT. Store at –20°C.
21. Klenow polymerase (5 U/µL). Store at –20°C.
22. 10 m*M* dATP in 10 m*M* Tris-HCl, pH 7.5. Store at –20°C.
23. 1 µg/µL Sheared salmon sperm DNA *(29)*. Store at –20°C.
24. Mouse 3T6 cells (American Type Culture Collection).
25. 100X Glutamine: 29 g L-glutamine in 1 L of water. Filter-sterilize and aliquot. Store at –20°C long term or at 4°C while in use.
26. 100X Pen/strep: 10,000 U (Gibco 15140-122). Store at –20°C long term or at 4°C while in use.
27. 1X Dulbecco's modified Eagle's medium (DMEM) (Gibco 12100-046): make according to manufacturer's directions. Add pen/strep to 1X and pH to 7.2 with HCl. Filter-sterilize through 0.2-µm filter and place one bottle at 37°C overnight to check for contamination. Add glutamine every 2 wk to 1X. Store at 4°C.

28. 60-mm Polystyrene tissue culture plates (Costar 0877221).
29. Heat-inactivated bovine calf serum (HICS) (Gibco 16170078): Thaw if frozen. Heat at 57°C for 1 h. Store at 4°C.
30. Newborn bovine serum (Gibco 16010159) (NBS). Store at 4°C if in use or at –20°C for long-term storage.
31. Glycerol.
32. 2X HEBS, pH 7.03 ± 0.05: per 200 mL add 3.2 g NaCl, 0.148 g KCl, 0.04 g Na$_2$HPO$_4$, 0.4 g D-glucose, and 2 g N-2-hydroxyethylpiperazine-N'-2-ethane-sulfonic acid (HEPES) to water. pH with NaOH and filter-sterilize. Store long term at –70°C in 10-mL aliquots. Store at 4°C while in use.
33. 1X HEBS: use same quantities as for 2X HEBS (**item 32**), but make up to 400 mL.
34. 20X PBS: for 1 L add 4 g KCl, 4 g KH$_2$PO$_4$, 160 g NaCl, 22.8 g Na$_2$HPO$_4$ to water. Autoclave and store at room temperature. (pH will be 7.2 when diluted to 1X.)
35. 2.5 M CaCl$_2$. Filter-sterilize and store at room temperature.
36. Sterile water.
37. Polystyrene tubes (Falcon 2054).
38. 1-mL Disposable polystyrene cotton-plugged pipets.
39. Stopwatch.
40. Cotton-plugged, sterile Pasteur pipets.
41. 4.5% NP40/4.5% Tween-20 in water. Store at room temperature.
42. Proteinase K (10 mg/mL): resuspend 10 mg in 1 mL TE and digest for 15–30 min at 37°C. Store at –20°C.
43. 95% Ethanol.
44. 3 M NaOAc, pH 6.0. Adjust pH with acetic acid and autoclave or filter-sterilize. Store at room temperature.
45. 24-Well polystyrene tissue culture plates (Corning 25820).
46. Rubber policeman.
47. 13-mL Polypropylene tubes with caps (Sarstedt 55.518 and 65.793).
48. Hirt solution 1: 10 mM Tris-HCl, pH 7.0, 10 mM EDTA. Store at room temperature.
49. Hirt solution 2: 10 mM Tris-HCl, pH 7.0, 10 mM EDTA, 1.2% sodium dodecyl sulfate (SDS), 2 mg/mL pronase added fresh before using. Store at room temperature.
50. 10% SDS: do not autoclave. Make with sterile water. Store at room temperature.
51. Pronase (20 mg/mL): prepare in water and incubate at 37°C for 2 h. Store in aliquots at –20°C.
52. 4-mL Polypropylene tubes with caps (Sarstedt 55.532 and 65.809).
53. 10X TBE: 108 g Tris base, 55 g boric acid, 40 mL 0.5 M EDTA, pH 8.0, per liter.
54. 0.7% Agarose in 1X TBE for gel electrophoresis.
55. 10X Taq polymerase buffer: 500 mM KCl, 100 mM Tris-HCl, pH 9.0, 1% Triton X-100. Store at -20°C.
56. 25 mM MgCl$_2$: use for PCR only and store at –20°C.

57. PCR primers: 250 ng/μL stocks of each. Store at –20°C.
58. Taq polymerase (5 U/μL). Store at –20°C.
59. 10 m*M* dNTPs: make stock in water from ultrapure dNTPs (Roche 1969064). Store at –20°C.
60. PCR tubes.
61. 7% Native polyacrylamide gel: 5 mL 10X TBE, 11.67 mL 30:0.8 acrylamide, 33.1 mL water. Mix by swirling and add 300 μL of 10% ammonium persulfate and 30 μL TEMED. Mix and cast immediately.
62. Ethidium bromide (10 mg/mL): dissolve in water and stir overnight in a container covered with foil. Store at room temperature wrapped in foil.
63. Wizard DNA Clean-up Kit (Promega 47280). Store at room temperature.
64. fmol Sequencing Kit (Promega Q4110). Store at room temperature.
65. γ-^{32}P dATP 3000 Ci/mmol. Store at –20°C.
66. 8% Urea, 6% polyacrylamide gel: 5 mL 10X TBE, 7.5 mL 38:2 acrylamide, 18.3 mL water, and 24 g urea. Stir for at least 15 min. Add 300 μL of 10% ammonium persulfate and 20 μL of TEMED and cast immediately.
67. Sequencing gel fix: 10% methanol/10% acetic acid. Store at room temperature in a brown bottle.
68. 35-mm Polystyrene tissue culture plates (Corning 25000-35 or 25810).
69. 100-mm Polystyrene tissue culture plates (Costar 0877222).
70. 150-mm Polystyrene tissue culture plates (Costar 0877224).
71. Small glass test tubes with caps.
72. 1.6% Agarose: weigh agarose (low EEO). Add water and autoclave. Store at room temperature.
73. 2X DMEM: for 2 L add 53.5 g DMEM powder (Gibco 12100-046), 14.8 g NaHCO$_3$. Adjust pH to 7.2 with HCl. Bring volume to 1700 mL. Filter-sterilize with 0.2-μm filter to 425-mL aliquots. Incubate one bottle at 37°C overnight to check for contamination. Per 425 mL of 2X DMEM, add 10 mL 100X glutamine, 10 mL 100X pen/strep, 10 mL 100X nonessential amino acids, 25 mL 1 *M* MgCl$_2$, and 20 mL HICS. When these components are added, the solution is 2X plaque assay medium. Add glutamine to 2X every 2 wk. Store at 4°C. Do not use if more than 2 mo old.
74. 100X Nonessential amino acids (Gibco 11140-050).
75. 1 *M* MgCl$_2$. Autoclave and store at room temperature. Use only for cell culture.
76. 100X Neutral red: 1% neutral red in water. Filter through Whatman no. 1, then through 0.2-μm filter. Store at room temperature.
77. TE-equilibrated phenol *(29)*.
78. Chloroform.
79. Oligonucleotide-directed mutagenesis kit (Amersham or Stratagene).
80. Plugged pipet tips.
81. StrataClean^{TM} resin (Stratagene 400714).

3. Methods

3.1. Preparation of DNA–Protein Complex

3.1.1. Collecting Virus

1. Obtain 1 L (approx 40 150-mm plates) fresh virus stock of MAV-1 (*see* **Subheading 3.4.**). To make mutants in the E1A or E3 regions, use wt variants pmE301 or pmE101, respectively (*see* **Note 1**).
2. Pour the virus stock into a 1-L graduated cylinder to determine the volume. Split the volume equally into two 1-L centrifuge bottles.
3. Add NaCl to 0.5 *M* using a 5 *M* NaCl stock (*see* **Note 2**).
4. Add 50% PEG to a final concentration of 8% based on the volume including the NaCl added in **step 3**.
5. Mix gently by inversion and incubate overnight on ice at 4°C.
6. Pellet virus by centrifugation at 2200g for 20 min at 4°C. It will be a white film that covers the bottom of the bottle.
7. Pour off the supernatant and discard after treating with bleach.
8. Add 6 mL of 10 m*M* Tris-HCl, pH 8.0, to each 1-L centrifuge bottle and gently resuspend by pipetting up and down. (Residual medium will be present in each bottle.) Incubate at 4°C for 30 min to overnight. Load onto CsCl step gradients.

3.1.2. Making a CsCl Step Gradient

1. Add 3.5 mL of 1.4 g/cm^3 CsCl to a 13-mL ultracentrifuge tube.
2. Use a 5-mL pipet to draw up approx 5 mL of 1.2 g/cm^3 CsCl, remove the bulb from the end of the pipet, and quickly cover with a gloveless hand, which allows for greater control of the rate of flow. Release the liquid into the stock container until 3.5 mL remains in the pipet.
3. Slowly layer the 1.2 g/cm^3 CsCl above the 1.4 g/cm^3 by dripping it down the side of the tube, holding the pipet close to the existing liquid level and keeping the tube at eye level to observe the formation of a layer between the two solutions. Continue dripping slowly until the gradient is complete. Make four gradients for this procedure (*see* **Note 3**).
4. Add up to 5–6 mL of virus (**Subheading 3.1.1.**, **step 8**) to each step gradient. If necessary, add 10 m*M* Tris-HCl, pH 8.0, so that the tube is filled to within a few millimeters of the top.
5. Centrifuge the samples in a Beckman SW41 swinging bucket rotor at 35,000 rpm (210,000g) for at least 90 min at 4–10°C (*see* **Note 4**). The density of adenovirus is 1.33–1.35.
6. After centrifugation, inspect the samples for two white bands. The lower, smaller band is virus and will be collected using a dripper (*see* **Note 5**). A black piece of paper behind the tubes will help to visualize the bands present after centrifugation.
7. Place a sample into the dripper and attach the top, making sure that the dripper needle is below the tube holder and will not immediately puncture the tube. Clamp the tubing with a hemostat or set up a peristaltic pump and turn on for a few seconds to put a small amount of positive pressure on the system. Turn the needle

until the tube is punctured and the liquid drips out briefly. Release the hemostat and control the flow manually or with the peristaltic pump. Begin to collect the lower white band, which contains virus, as it nears the bottom of the tube. It will appear white or translucent against a black paper background. The remaining material, including the upper band, has incomplete capsids, cellular DNA, and lipids and can be discarded. Repeat for each sample.

8. Dilute each sample with one volume of 10 mM Tris-HCl, pH 8.0, to get $\rho \leq 1.2$ and load onto a linear gradient.

3.1.3. Making a CsCl Linear Gradient

1. Attach 1-mm inner-diameter tubing, long enough to be used with a peristaltic pump, to the outlet valve on the gradient maker, and insert a capillary tube into the free end of the tube. Adjust the flow rate to approx 1 mL/min.
2. Close the outlet valve on the gradient maker and add 5 mL of 1.2 g/cm^3 CsCl and a stir bar to the chamber proximal to the outlet valve. Open the valve between the two chambers and allow a small amount of the liquid to flow into the distal chamber to remove the air from the valve. Close the valve and pipet the liquid from the distal chamber to the proximal chamber. Add 5 mL of 1.4 g/cm^3 CsCl to the distal chamber.
3. Place the tubing into the bottom of a 13-mL ultracentrifuge tube. Start the stirrer, then open both valves and turn on the peristaltic pump simultaneously. Fill the tube from the bottom and monitor the progress of the gradient so that no bubbles are released as the gradient is completed. Gently remove the capillary tube from the solution by sliding it up one wall of the tube so as not to disturb the gradient. Make four to six gradients for this procedure.
4. Load the diluted virus from **Subheading 3.1.2.**, **step 8,** to the top of the gradients. As necessary, add 10 mM Tris-HCl, pH 8.0, to fill tubes to within a few millimeters of the top of the tube.
5. Centrifuge at 210,000g in an SW41 swinging bucket rotor at 4°C for at least 16 h (*see* **Note 4**).
6. In the second gradient a major band should be seen, with possibly a faint band above it. Using the dripper as described above, collect the bottom white band, which contains virus. Discard the rest of the gradient, including the top white band containing mostly cellular debris and top components.
7. Dialyze the samples for 2 h to overnight against 10 mM Tris-HCl, pH 8.0, at 4°C in 12,000–14,000 molecular-weight exclusion dialysis tubing (*see* **Note 6**).
8. Place no more than 3.5 mL of sample in a conical polypropylene 15-mL tube with graduations and add the following: 70 μL 1 M Tris-HCl, pH 8.0, 14 μL 0.5 M EDTA, 3.5 mL 8 M guanidine HCl, 3.57 g optical grade CsCl, and dH$_2$O to a final volume of 7 mL (*see* **Note 7**).
9. Mix well by inversion and pour into a 13-mL ultracentrifuge tube. Fill to within a few millimeters of the top with mineral oil.
10. Centrifuge in an SW41 swinging bucket rotor at 28,000 rpm (134,000g) for 40 h at 15°C (*see* **Note 4**).

11. Using the dripper and the peristaltic pump as described above, fractionate the gradient into 0.5-mL aliquots (~20–25 drops per fraction), collecting 15–20 fractions per gradient in capless microfuge tubes.

12. Measure the A_{260} of each fraction (dilute fractions in TE if necessary) and plot. Pool the fractions that form the peak.

13. Dialyze in 12,000–14,000 molecular-weight exclusion dialysis tubing against 2 L of chilled TE at 4°C for four changes of at least 4 h each.

14. Measure A_{260} and calculate the concentration using the following formula: (A_{260}) (50 µg/mL) (dilution factor)/1000 = concentration in µg/µL. The total yield should be 100–200 µg).

15. Store in microfuge tubes in 5- or 10-µg aliquots. Freeze quickly in dry ice or liquid nitrogen and store at –70°C.

3.1.4. Preparing DNA–Protein Complex for Transfection

1. Thaw 5 µg of DNA–protein complex (**Subheading 3.1.3.**, **step 15**) per transfection sample and digest it with 8 µL of *Eco*RI (20 U/µL) in a final concentration of 1X *Eco*RI buffer containing 1 m*M* DTT in a final volume of 250 µL. Digest 4 h to overnight at 37°C. (*See* **Subheading 3.2.** to determine the number of aliquots of DNA–protein complex needed.)

2. Partially fill in the sticky ends created by the *Eco*RI using 8 U of Klenow polymerase and a final concentration of 1 m*M* dATP. Incubate for 30 min at 37°C. Stop the reaction by adding EDTA to a final concentration of 25 m*M*. Store at –20°C until ready to use.

3. Linearize 4 µg of plasmid containing the desired MAV-1 mutation(s) by cutting at a unique site within the vector. Store at –20°C until ready to use (*see* **Note 8**).

3.2. Transfection of Mouse Cells

1. For each transfection, it is important to include the following controls. As a negative control (no. 4, **Table 1**), add only salmon sperm DNA to the transfection mix. As a positive control (no. 3, **Table 1**), transfect undigested complex alone. As a control for background levels of religation of digested, partially filled-in DNA-protein complex (no. 2, **Table 1**), include digested, partially filled-in complex alone. As a positive control for the plaque assay, set up a plaque assay with a stock of the virus used to make the DNA–protein complex.

2. One or 2 d prior to transfection, pass mouse 3T6 cells to 60-mm tissue culture plates in 1X DMEM containing 5% heat-inactivated bovine calf serum. The cells should be approx 70% confluent at the time of transfection. (There will be approx 1×10^7 mouse 3T6 cells on a 100-mm plate when confluent.) Set up two plates for each transfection that is planned, i.e., two plates for each control listed above (**Subheading 3.2.**, **step 1**) and two plates for each mutant (*see* **Note 9**).

3. On the day of the transfection, warm the following to 37°C: 3 mL/transfection plate of 1X DMEM containing 5% HICS, 5 mL/transfection and plaque assay plate of 1X DMEM containing 2% newborn bovine serum (NBS), 1.5 mL/trans-

Table 1
Transfection Experimental Plan

Tube no.	DNA–protein complex		Mutant plasmid		salmon sperm DNA, µL	2.5 M CaCl$_2$, µL	H$_2$O to 500 µL
	µg	µL	µg	µL			
1	5 (digested, filled in)	290.5	4	10	11	50	138.5
2	5 (digested, filled in)	290.5	—	—	15	50	144.5
3	5 (undigested)	290.5	—	—	15	50	144.5
4	—	—	—	—	20	50	430.0

    fection plate of 15% glycerol in 1X DMEM, 1.5 mL/transfection plate of 1X HEBS, pH 7.03 ± 0.05, and approx 5 mL/transfection plate of 1X PBS (*see* **Note 10**).

4. Have the following at room temperature: 0.5 mL per transfection of 2X HEBS, pH 7.03 ± 0.05, 50 µL per transfection of 2.5 M CaCl$_2$, and up to 500 µL of sterile water per transfection.

5. Thaw the following and store on ice until ready to use: 5 µg of digested, filled-in DNA–protein complex per experimental transfection and background control, 5 µg of undigested DNA–protein complex per positive control, 4 µg of linearized mutant plasmid per experimental transfection, and salmon sperm DNA (1 µg/µL).

6. Pipet 0.5 mL of room temperature 2X HEBS, pH 7.03 ± 0.05, into a numbered polystyrene tube (Falcon 2054) for each pair of plates to be transfected.

7. Set up a DNA mix for each transfection in a 1.5-mL microfuge tube as in **Table 1**. Using salmon sperm DNA (1 µg/µL), adjust the total amount of DNA to 20 µg and using sterile water adjust the final volume to 500 µL. Numbers are as follows: 1, experimental sample; 2, background religation control; 3, positive control; 4, negative control.

8. Vortex each of the transfection mix tubes and proceed immediately to the subsequent steps.

9. Aspirate medium from two to four plates at a time, and add 1.5 mL of warm 1X HEBS, pH 7.03 ± 0.05 to each plate. Number the plates as above, remembering that the calcium phosphate precipitate will be split between two plates, generating plates 1A and 1B, and so on.

10. Using a 1-mL disposable plugged polystyrene pipet, add the DNA mix for the first sample to the polystyrene tube (prepared in **step 6**). Do this slowly over approx 45 s, while gently flicking (*see* **Note 11**). As soon as a precipitate is seen, add half of the mix to each of the appropriately labeled plates containing 1.5 mL of 1X HEBS, pH 7.03 ± 0.05 (prepared in **step 9**). Start a stopwatch and note the

time. Let the plates sit for 20 min at room temperature.

11. Repeat **steps 9** and **10** for the rest of the samples, allowing 3–5 min between each sample.

12. At the end of the 20-min incubation, add 3 mL of warmed 1X DMEM containing 5% HICS. Check each plate under the microscope for a sandy, grainy appearance of the precipitate. (If the precipitate appears clumpy, the transfection may not work.) Incubate at 37°C.

13. After incubating for 3–4 h, glycerol-shock the cells. Remove the medium from two plates at a time. Add 1.5 mL of warmed 15% glycerol in 1X DMEM and let it sit for exactly 60 s (use a stopwatch). Aspirate the medium containing glycerol and wash carefully with 3–5 mL of warmed 1X PBS. Aspirate the PBS and add 5 mL 1X DMEM containing 2% NBS.

14. After completing the transfection, infect the remaining cells for a plaque assay with the parent virus from which the DNA–protein complex was made. Follow the directions in **Subheading 3.5.** to infect the cells for a plaque assay, but instead of overlaying with agarose, add 5 mL/plate of 1X DMEM containing 2% NBS after infecting. Incubate overnight at 37°C.

15. The next day, aspirate the medium from one set of the transfection plates from **step 13** (i.e., 1A, 2A, etc.) and overlay with 5 mL of medium/agarose as in **Subheading 3.5.** Change the medium on the duplicate plates (1B, 2B, etc.) to fresh 1X DMEM containing 5% HICS. Overlay all of the plates for the virus infection (plaque assay control) from **step 14**. Follow the directions for overlaying and checking for plaques as in **Subheading 3.5.**

3.3. Isolation and Identification of Mutants

3.3.1. Picking Plaques and Extracting Viral DNA

1. After plaques have formed on the experimental plates, circle plaques, using an ethanol-soluble marker, and use a plugged Pasteur pipet to pick each of the plaques from the experimental plates first. Then pick one or two plaques from the positive control plate or the background plate. Pipet the agarose plug into 0.5 mL of 1X PBS. Vortex the samples to break up the agarose and distribute the virus. **Note**: In all the following steps, be careful not to cross-contaminate samples because they will be used in a PCR assay (*see* **Note 12**).

2. Using a plugged pipet tip, aliquot one-quarter to one-half of each of the plaque-containing solutions to PCR clean 1.5-mL microfuge tubes. Freeze the remaining portions at –20°C.

3. Add approx 20 ng of purified MAV-1 DNA (*see* **Note 6**) to a 1.5-mL microfuge tube as a control for the DNA extraction procedure.

4. To extract the DNA from the plaque solution, add 0.1 volume of 4.5% NP40/ 4.5% Tween-20 and 250 µg/mL proteinase K. Digest at 55°C for 1 h. Incubate at 95°C for 10 min to inactivate the proteinase K.

5. Ethanol-precipitate the DNA by adding 0.1 volume of 3 *M* NaOAc, pH 6.0, and 2.5 vol of 95% ethanol. Incubate on ice or at –20°C for 15 min. Centrifuge for 15

min at 4°C or room temperature. Discard the supernatant and air-dry the pellet. (A pellet may or may not be visible.)

6. Resuspend the pellet in 10 µL of sterile water and use 1–5 µL in a PCR reaction.

3.3.2. Hirt Method for Obtaining Viral DNA From Plaques (28) (see **Note 13**)

1. Infect one well of a 24-well plate using one-quarter to one-half of each of the solutions derived from a plaque. Follow the protocol found in **Subheading 3.4.1.**
2. Monitor the cells daily for cytopathic effect (CPE) (*see* **Note 14**). Harvest the cells when CPE is visible by scraping the cells into the medium using a rubber policeman. Freeze–thaw the stock three times, and centrifuge to remove the cellular debris. Use this stock to infect one to two 60-mm plates of mouse 3T6 cells that are 75–80% confluent. When early signs of CPE are evident, such as a slight rounding of cells and some refractile cells (~48 h postinfection), wash the monolayer once with 1X PBS. Scrape cells with a rubber policeman into 1X PBS and transfer the solution to a 13-mL polypropylene tube. Centrifuge at 2000–3000*g* for 10 min at 4°C.
3. Aspirate the supernatant without disturbing the cell pellet. (The cell pellets can be frozen at –20°C if necessary.)
4. Resuspend each pellet in 0.5 mL of Hirt solution 1 and vortex very gently to resuspend the cells.
5. While gently vortexing, add 0.5 mL of Hirt solution 2 to which 0.1 volume of 20 mg/mL pronase has been added.
6. Cap the tubes and incubate at 37°C for 2 h.
7. Add 0.25 mL of 5 *M* NaCl to each sample and flick the bottom of each tube to mix. Incubate overnight on ice at 4°C.
8. Centrifuge at 17,000*g* for 45 min at 4°C.
9. Transfer supernatant to a 4-mL polypropylene tube and extract twice with an equal volume of phenol.
10. Add 2.5 volumes of ethanol to each tube and incubate on ice or at –20°C for 15 min. Centrifuge at 10,000*g* for 15 min at 4°C. Discard the supernatant. (No salt is needed to precipitate the DNA because the supernatant contains 1 *M* NaCl.)
11. Resuspend pellet in 0.3 *M* NaOAc, pH 6.0, and transfer to a 1.5-mL microfuge tube. Add 1 mL 95% ethanol to tube and precipitate as above.
12. Wash pellet with 1 mL of 70% ethanol and centrifuge for 5 min.
13. Air-dry the pellet and resuspend in 50 µL of water or TE.
14. Use 1–5 µL in PCR or 10 µL per restriction enzyme digest and electrophorese on a 0.7% agarose gel. If using in a restriction digest, also add 1 µg RNase per sample. Include appropriate control DNA samples in analysis.

3.3.3. PCR Amplification of Potential Mutants

1. The following PCR reaction components can be used to amplify DNA obtained in **Subheading 3.3.1.** or **3.3.2.** PCR reactions should be set up with the appropri-

ate positive and negative controls. Standard PCR conditions include 1X *Taq* polymerase buffer, 1 m*M* MgCl$_2$, 0.2 m*M* dNTPs, 250 ng of each primer, 1 U of *Taq* polymerase, 1–5 μL of viral template from **Subheading 3.3.1.** or **3.3.2.** in a volume of 25 μL. 0.1–1.0 ng of MAV-1 DNA or 0.05–0.1 ng of the plasmid containing the wt DNA sequence or the mutation of interest can be used as a positive control.

2. Use the following PCR conditions for machines that accommodate 0.2-mL tubes. Denature at 94°C for 2 min. Repeat the following program for 40 cycles: denature for 15 s at 94°C, anneal for 15 s at 44°C, and extend for 45 s at 72°C. Finally, extend at 72°C for 3 min followed by 4°C indefinitely. (If using a machine that accommodates 0.5-mL tubes, the times for each step may need to be adjusted.)

3.3.4. Analysis of PCR Products for Potential Mutants

1. Once PCR amplification has been completed, the mutants can be screened directly by electrophoresis, by restriction digestion and electrophoresis, or by sequencing.
2. If the mutation results in a significantly smaller PCR product than that of wt (i.e., the mutation is a deletion), the completed PCR reactions may be simply electrophoresed on a 7% native polyacrylamide gel after loading dye has been added (5% glycerol final concentration) *(1)*. Electrophorese at 200 V for 2 h. Stain with ethidium bromide for 10–20 min and destain with water for 10–20 min. View and photograph on a UV transilluminator.
3. If a restriction site that is unique to the PCR product has been incorporated into the mutation, the PCR product can be digested with the appropriate enzyme to determine if the DNA is mutated *(1)*. In this case, aliquot half of each completed PCR reaction to microfuge tubes and digest with the appropriate enzyme. Electrophorese the undigested and digested samples using electrophoresis and detection conditions as described above.
4. Point mutations may also be screened using differential PCR *(4)*. In this case two primers are designed that will differentially amplify the wt and mutant DNAs. This can be accomplished by having the 3' nucleotide in each oligonucleotide correspond to the wt or mutant sequence. The PCR conditions to differentially amplify mutant and wt products will have to be empirically determined. To do this, try varying the Mg^{+2} concentration and/or the annealing temperature. The PCR products are then electrophoresed and detected as described.
5. If the mutation does not meet the above criteria, the PCR product can be sequenced. After the PCR reaction is complete, remove the excess primers using the Wizard DNA Clean-up Kit (Promega). Analyze 5–10 μL of the cleaned PCR product by electrophoresis to ensure that the primers have been removed and that the expected PCR product was produced. Use 5 μL of the purified PCR product as template in sequencing reactions, following the directions given in Promega's fmol Sequencing Kit. The best results are obtained using end-labeled ^{32}P-oligonucleotides as the sequencing primers. Using the bromophenol blue and xylene cyanol dyes as markers for oligonucleotide migration, electrophorese the sequencing reactions on an 8% urea, 6% polyacrylamide gel until the area of

interest is in a readable position on the gel. Fix the gel in 10% methanol/10% acetic acid for 30–45 min and rinse in water briefly. Dry on a gel dryer and expose to film overnight. Read the film after exposure to determine if the desired mutations have been incorporated into the viral DNA.

3.3.5. Purification of Mutant Plaques

1. Once the mutations of interest have been identified by one of the methods in **Subheading 3.3.4.**, the mutant virus should be plaque-purified to ensure that the virus stock is not contaminated with wt virus.
2. To plaque-purify the mutants, set up one to two 60-mm plates of mouse 3T6 cells (*see* **Subheading 3.2.**, **step 2**) per mutant obtained in the above screen. The cells should be approx 70% confluent at the time of infection.
3. Aspirate medium from the cells and infect with 0.125–0.25 mL of the solution of virus derived from a plaque (*see* **Subheading 3.3.1.**, **step 2**). Incubate for 1 h at 37°C. Following incubation, overlay with 5 mL of medium/agarose and maintain for 10–14 d as described in **Subheading 3.5.**
4. When plaques appear, pick (10 per plate if available) and screen them as before.
5. Choose a plaque from each plate that appears to be free of most of the background wt virus and repeat the plaque purification procedure. If after this third round of plaque purification there is still some contamination by wt virus, repeat the purification again. If not, prepare a virus stock of the mutants from the most purified plaque (*see* **Subheading 3.4.**).

3.4. Preparation of Virus Stocks

3.4.1. Making a Virus Stock From a Plaque

1. Set up one to two wells (24-well plate) of mouse 3T6 cells per purified mutant plaque so that they will be 75–80% confluent at the time of infection.
2. When the cells are ready, infect one or two wells per mutant by adding one-quarter to one-half of the purified plaque stock (*see* **Note 15**). Incubate at 37°C for 1 h and add approx 1 mL of prewarmed 1X DMEM containing 1% HICS. Monitor for CPE (*see* **Note 14**).
3. When the cells have significant CPE or have been infected for 10–14 d, harvest them by scraping the cells into the medium with a rubber policeman (*see* **Note 16**). Freeze–thaw the virus three times, then centrifuge 5 min at 2200*g* to remove the cellular debris. Store the virus at –70°C.
4. Using the newly obtained virus stocks and following the directions given above, infect one 35-mm plate of 75–80% confluent mouse 3T6 cells per mutant (*see* **Note 17**). Harvest as above and repeat using the new virus stock to infect the next larger plate of cells (35-mm→60-mm→100-mm→150-mm). Because the titer of the virus stocks is unknown, estimate the amount of virus needed. As a starting point, use approx 0.5–1 mL to infect 35-mm plates, 1 mL to infect 60-mm plates, 2 mL to infect 100-mm plates, and 3 mL to infect 150-mm plates. (Two 150-mm plates of cells will need to be infected if a large virus stock of greater than 200 mL is to be generated.)

3.4.2. Making a Large Virus Stock

1. Set up 10 150-mm plates of 3T6 cells per mutant so that they are 75–80% confluent at the time of infection.
2. Aspirate the medium from each plate and add approx 3 mL of mutant virus per plate. Rock the plates to ensure that all the cells are covered by the liquid. Incubate at 37°C for 1 h, then add 22 mL of warmed 1X DMEM containing 1% HICS. Monitor the cells for CPE and harvest when most of the cells are dead and detached from the plate. Harvest the virus by collecting the medium and spinning out the cells or by scraping the cells into the medium, freeze–thawing three times, and centrifuging to remove the cellular debris (*see* **Note 16**). Aliquot approx 1 mL of each mutant stock to a tube to use in a plaque assay. Store the remainder at –70°C in a large volume or as smaller aliquots for easy thawing upon use.

3.5. Quantitation of Virus by Plaque Assay

1. For each mutant, set up 13 60-mm plates of 3T6 cells to be approx 70% confluent at the time of infection.
2. For each mutant, aliquot exactly 0.9 mL of sterile PBS to seven small glass test tubes with caps and number and order these tubes from 2 to 8.
3. Warm to 37°C the 1-mL aliquot of each virus stock to be titrated (**Subheading 3.4.2., step 2**). Do not leave at 37°C for prolonged periods. Once the stock is thawed, store on ice.
4. Aliquot exactly 0.1 mL of virus into tube 2. Vortex and remove exactly 0.1 mL from tube 2, and deposit it into tube 3. Vortex and repeat for the remaining tubes (changing tips each time), creating a dilution series of the original virus stock. Repeat for each virus stock.
5. Label two plates per dilution (10^3–10^8) and one plate as mock. Remove medium from plates before adding PBS or virus solution. Add 0.1 mL of PBS to the mock plate.
6. Beginning with the highest tube number (10^8), the most dilute virus sample, add exactly 0.1 mL of the solution to each of the two plates labeled 10^8. Rock the plates to cover the cells with the liquid. Repeat sequentially in descending order, but omitting tube 2, as there will be too many plaques to count from this dilution (*see* **Note 18**). If you have worked from most dilute to least dilute, as described above, you do not have to change tips with each sample.
7. Incubate at 37°C for 1 h. Meanwhile, microwave sterile 1.6% agarose until it is boiling and place it in a 45°C water bath. Allow agarose to equilibrate at 45°C for at least 30 min. (You will need 2.5 mL per plate, but allow some extra.) Warm 2X plaque assay medium containing 4% HICS that is supplemented with glutamine, pen/strep, and nonessential amino acids all at a final concentration of 2X, as well as $MgCl_2$ that is at a final concentration of 50 mM (*see* **Note 19**). (You will need 2.5 mL per plate, but allow some extra.)
8. After the 1 h incubation, mix equal volumes of warm 2X DMEM and agarose. Working quickly, overlay each plate with 5 mL of medium/agarose by gently applying the solution to the cells with a wide-mouth pipet by adding to the side of

the plate, rather than dispensing the medium to the center of the plate. Try not to introduce bubbles onto the plates. Allow to stand about 30 min. Place the plates in the 37°C incubator.

9. The next day, and every second or third day thereafter, overlay each plate as described above with 2 mL per plate, i.e., 1 mL of 2X plaque assay medium mixed with 1 mL of agarose.

10. When plaques become visible (days 8–9), begin counting and recording the number of plaques. Continue counting daily and overlaying until no or only a few new plaques appear for 1–2 d (*see* **Note 20**). Always count the plaques prior to overlaying because the recent addition of medium/agarose tends to make the plaques very hard to see. If desired for added visibility of plaques, add neutral red to a final concentration of 1X to the medium each time the plates are overlaid once plaques are first visible.

11. When no or very few new plaques appear for 1–2 d, total the number of plaques for each plate and multiply that number by the dilution factor. This number is the titer of the virus in PFU/mL. The most accurate titer will be determined from at least two plates with 10–100 plaques per plate. For example, if two 10^6 plates yielded a total of 47 and 53 plaques, respectively, the average of 50 plaques is multiplied by the dilution factor (10^6) to yield a titer of 5×10^7 PFU/mL. In some cases titers from two dilutions can be averaged to give the titer of the virus stock (*see* **Note 21**).

4. Notes

1. *pm*E301 and *pm*E101 are MAV-1 wt variants that contain only one *Eco*RI site in the genome. *pm*E301 contains a single *Eco*RI site in the E1A region, whereas *pm*E101 contains a single site in the E3 region (*1,4*).

2. 1X DMEM contains 0.15 *M* salt, and because the volume of NaCl contributed from the 1X DMEM is not negligible, be sure to include it in the calculation.

3. Keep movement of the gradient tubes to a minimum to avoid disturbing the gradients. Make sure that a layer is visible in each one prior to loading your samples. If no layer is present, discard and make another gradient.

4. Because these are equilibrium density gradients, the centrifugation can go longer than the indicated time.

5. We use a Hoeffer dripper: follow the manufacturer's instructions for your dripper. Using the dripper can be tricky, so it is advisable to practice the procedure on a blank tube first. This allows you to work out the best way to puncture the tube and to control the flow rate. We also recommend cleaning the dripper needle after each gradient to prevent clogging. To clean, insert a needle through the opening a few times to free any plastic. The dripper needle should be washed by injecting ethanol and/or water into the opening with a syringe. We recommended cleaning the needle after every sample or every other sample and when you finish a procedure.

6. Purified MAV-1 DNA can be made by digesting the dialyzed virus in 0.5% SDS and 250 µg/mL of proteinase K at 50–55°C for 15 min. Extract with phenol one

or two times and re-extract the first phenol phase with TE. Extract with chloroform twice to remove the phenol. Dialyze against TE, pH 8.0, at 4°C for four changes of at least 4 h each. Read the A_{260} and store at 4°C. Expect the yield to be ≤0.1 mg/mL.

7. When making the gradient containing guanidine HCl, put the dialyzed virus onto two gradients to have a balance in the centrifuge.

8. To generate a plasmid with a mutation in an MAV-1 gene, first clone the wt gene of interest into a vector and mutagenize using oligonucleotide-directed mutagenesis (Amersham or Stratagene), PCR mutagenesis, or by replacing genomic DNA sequence with cDNA sequence. To screen for potential mutants generated by oligonucleotide-directed mutagenesis, a unique restriction site. Mutagenesis is a unique restriction site that can be incorporated into the mutation. The presence of the site will distinguish mutant from wt in subsequent steps. To prevent digestion of the plasmid DNA by the enzyme used to digest DNA–protein complex (when the two are mixed together for the transfection), you can first treat the plasmid DNA with the appropriate methylase (*Eco*RI methylase in this case), following the manufacturer's instructions *(4)*.

9. A complementing cell line may be necessary to obtain a stock of some mutants.

10. Other transfection methods may also be used, but this method has been most consistently successful in our experience.

11. Alternatively, air can be bubbled into the 2X HEBS while the DNA mix is added.

12. Use general precautions to prevent contamination of samples that will be used for PCR. For example, always use plugged pipet tips and remove clean tubes and reagents from containers with forceps or with newly gloved hands that have not come in contact with any DNA. Also, use water that is sterile and used only for PCR to make solutions used for this procedure or in the PCR reaction itself.

13. DNA extraction methods of Shinagawa et al. *(30)* have also been successfully used to isolate MAV-1 DNA from infected cells (A. Kajon and K. Spindler, unpublished).

14. When inoculating cells with a plaque, the viral titer may be very low; thus, CPE may not begin to appear for up to 10–14 d. However, the cells should be checked daily to determine if CPE is present. Harvest the virus-infected cells after 10–14 d if no CPE is visible and use as a virus stock.

15. Do not use all of the purified plaque stock in case you need to repeat the procedure. Also, do not set up a wt virus infection as a control when growing up stocks of the mutant viruses because it will greatly increase the chance of the mutant stocks becoming contaminated with wt virus.

16. MAV-1 is released into the medium in a wt infection; thus only the medium is usually harvested to make a virus stock. However, the effect of the mutation(s) is unknown, so scraping the cells into the medium and freeze–thawing is recommended for mutants until it can be determined that the virus is released from the cells as in a wt infection.

17. Once there is a virus stock available, the DNA sequence of the virus should be confirmed to determine if the desired mutations are present and if the stock is pure. DNA for sequencing may be obtained from 60-mm plates using the Hirt

method described in **Subheading 3.3.2.** We have been unable to sequence directly from these DNAs; therefore, we recommend PCR amplifying the regions of interest, removing the primers, and sequencing using the fmol Sequencing Kit (Promega) (*see* **Subheading 3.3.4., step 5**). Alternatively, the DNA may be easily obtained by the following protocol. Remove 100 µL of medium from cells that are in the late stages of infection and boil for 5 min to denature the viral structural proteins. Centrifuge the samples for 10 s. Mix the stock container of StrataClean™ resin (Stratagene) and add 10 µL of resin to each sample to bind the proteins. Flick the tube to mix and centrifuge for 30 s to pellet the resin. Transfer the supernatant to a clean tube and use 1–2 µL for PCR. Follow **Subheadings 3.3.3.** and **3.3.4.** to PCR amplify and sequence these DNAs.

18. The tubes are numbered according to what their dilution on the plate (DOP) will be. The dilution in the first tube (no.2) is a 10-fold dilution. Only 0.1 mL of this will be used to inoculate the plate, and thus it is a 100-fold dilution; the DOP will be 10^2.

19. When overlaying cells with medium/agarose for a plaque assay, keep the medium and agarose separate until ready to overlay. Warm the medium in a container that is big enough to accommodate the agarose as well. When ready to overlay, add an equal volume of agarose to the aliquot of medium and use immediately. Have separate aliquots of medium/agarose for each mutant when overlaying and overlay the mock plate first, followed by the plates infected with the least amount of virus (10^8), continuing in order to those plates infected with the most amount of virus (10^3). Prepare enough medium/agarose to overlay one to two extra plates.

20. The plaques should be visible as holes in the monolayer of cells or 1- to 2-mm circular areas where the density of the monolayer appears lighter or heavier than the surrounding area when the plates are held up to the light. The formation of a plaque can be confirmed by circling the area with an ethanol soluble pen and viewing under the microscope using the 10X objective. The plaque will appear as a roughly circular area of dead cells. The plaques will be more difficult to see than human adenovirus plaques, and seeing them macroscopically is easier if observed against a black background. We have a piece of black paper on the ceiling near an overhead light and this seems to be crucial for seeing MAV-1 plaques.

21. Wild-type virus generally grows to a titer of approx 10^7 or 10^8 PFU/mL; however, some mutants, depending on the nature of the defect, may grow to 10- to 100-fold lower titers than wt virus. The titer of the stocks may also decrease on multiple freeze–thaws, so we recommend aliquotting the large stock of virus, especially if small amounts of virus are needed to obtain the desired multiplicity of infection.

References

1. Beard, C. W. and Spindler K. R. (1996) Analysis of early region 3 mutants of mouse adenovirus type 1. *J. Virol.* **70,** 5867–5874.
2. Cauthen, A. N., Brown, C. C., and Spindler, K. R. (1999) In vitro and in vivo

characterization of a mouse adenovirus type 1 early region 3 mutant. *J. Virol.* **73,** 8640–8646.

3. Cauthen, A. N. and Spindler, K. R. (1999) Novel expression of mouse adenovirus type 1 early region 3 gp11K at late times after infection. *Virology* **259,** 119–128.

4. Smith, K., Ying, B., Ball, A. O., Beard, C. W., and Spindler, K. R. (1996) Interaction of mouse adenovirus type 1 early region 1A protein with cellular proteins pRb and p107. *Virology* **224,** 184–197.

5. Kajon, A. E. and Spindler, K. R. (2000) Mouse adenovirus type 1 replication *in vitro* is resistant to interferon. *Virology* **274,** 213–219.

6. Fang, L., Stevens, J. L., Berk, A. J., and Spindler, K. R. (2004) Requirement of Sur2 for efficient replication of mouse adenovirus type 1. *J. Virol.* **78,** 12,888–12,900.

7. Smith, K. and Spindler, K. R. (1999) Murine adenovirus, in *Persistent Viral Infections* (Ahmed, R. and Chen, I., eds.), John Wiley & Sons, New York, pp. 477–484.

8. Ball, A. O., Beard, C. W., Villegas, P., and Spindler, K. R. (1991) Early region 4 sequence and biological comparison of two isolates of mouse adenovirus type 1. *Virology* **180,** 257–265.

9. Ishibashi, M. and Yasue, H. (1984) Adenoviruses of animals, in *The Adenoviruses* (Ginsberg, H. S., ed.), Plenum Press, New York, pp. 497–562.

10. Pirofski, L., Horwitz, M. S., Scharff, M. D., and Factor, S. M. (1991) Murine adenovirus infection of SCID mice induces hepatic lesions that resemble human Reye syndrome. *Proc. Natl. Acad. Sci. USA* **88,** 4358–4362.

11. Guida, J. D., Fejer, G., Pirofski, L.-A., Brosnan, C. F., and Horwitz, M. S. (1995) Mouse adenovirus type 1 causes a fatal hemorrhagic encephalomyelitis in adult C57BL/6 but not BALB/c mice. *J. Virol.* **69,** 7674–7681.

12. Kring, S. C., King, C. S., and Spindler, K. R. (1995) Susceptibility and signs associated with mouse adenovirus type 1 infection of adult outbred Swiss mice. *J. Virol.* **69,** 8084–8088.

13. Winters, A. L., Brown, H. K., and Carlson, J. K. (1981) Interstitial pneumonia induced by a plaque-type variant of mouse adenovirus. *Proc. Soc. Exp. Biol. Med.* **167,** 359–364.

14. van der Veen, J. and Mes, A. (1973) Experimental infection with mouse adenovirus in adult mice. *Arch. Gesamte Virusforsch.* **42,** 235–241.

15. Spindler, K. R., Moore, M. L., and Cauthen, A. N. (2006) Mouse adenoviruses, in *The Mouse in Biomedical Research,* 2nd ed., Vol. II, Academic Press, New York.

16. Meissner, J. D., Hirsch, G. N., LaRue, E. A., Fulcher, R. A., and Spindler, K. R. (1997) Completion of the DNA sequence of mouse adenovirus type 1: Sequence of E2B, L1, and L2 (18–51 map units). *Virus Res.* **51,** 53–64.

17. Larsen, S. H. and Nathans, D. (1977) Mouse adenovirus: growth of plaque-purified FL virus in cell lines and characterization of viral DNA. *Virology* **82,** 182–195.

18. Temple, M., Antoine, G., Delius, H., Stahl, S., and Winnacker, E.-L. (1981) Replication of mouse adenovirus strain FL DNA. *Virology* **109,** 1–12.

19. Chinnadurai, G., Chinnadurai, S., and Brusca, J. (1979) Physical mapping of a large-plaque mutation of adenovirus type 2. *J. Virol.* **32,** 623–628.
20. Kapoor, Q. S. and Chinnadurai, G. (1981) Method for introducing site-specific mutations into adenovirus 2 genome: construction of a small deletion mutant in VA-RNA$_I$ gene. *Proc. Natl. Acad. Sci. USA* **78,** 2184–2188.
21. Stow, N. D. (1981) Cloning of a DNA fragment from the left-hand terminus of the adenovirus type 2 genome and its use in site-directed mutagenesis. *J. Virol.* **37,** 171–180.
22. Nguyen, T. T., Nery, J. P., Joseph, S., et al. (1999) Mouse adenovirus (MAV-1) expression in primary human endothelial cells and generation of a full-length infectious plasmid. *Gene Ther.* **6,** 1291–1297.
23. Wigand, R., Gelderblom, H., and Özel, M. (1977) Biological and biophysical characteristics of mouse adenovirus, strain FL. *Arch. Virol.* **54,** 131–142.
24. Larsen, S. H. (1982) Evolutionary variants of mouse adenovirus containing cellular DNA sequences. *Virology* **116,** 573–580.
25. Pettersson, U. and Sambrook, J. (1973) Amount of viral DNA in the genome of cells transformed by adenovirus type 2. *J. Mol. Biol.* **73,** 125–130.
26. Dunsworth-Browne, M., Schell, R. E., and Berk, A. J. (1980) Adenovirus terminal protein protects single stranded DNA from digestion by a cellular exonuclease. *Nucleic Acids Res.* **8,** 543–554.
27. Gorman, C. (1985) High efficiency gene transfer into mammalian cells, in *DNA Cloning: A Practical Approach* (Glover, D. M., ed.), IRL Press, Oxford, pp. 143–190.
28. Hirt, B. (1967) Selective extraction of polyoma DNA from infected mouse cell cultures. *J. Mol. Biol.* **26,** 365–369.
29. Sambrook, J., Fritsch, E. F., and Maniatis, T., eds. (1989) *Molecular Cloning: A Laboratory Manual,* 2nd ed. Cold Spring Harbor Laboratory Press, Cold Spring Harbor, NY.
30. Shinagawa, M., Matsuda, A., Ishiyama, T., Goto, H. and Sato, G. (1983) A rapid and simple method for preparation of adenovirus DNA from infected cells. *Microbiol. Immunol.* **27,** 817–822.

5

Generation of Recombinant Adenovirus Using the *Escherichia coli* BJ5183 Recombination System

P. Seshidhar Reddy, Shanthi Ganesh, Lynda Hawkins, and Neeraja Idamakanti

Summary

One of the most time-consuming steps in the generation of adenoviral vectors is the construction of recombinant plasmids. This chapter describes a detailed method for the rapid construction of adenoviral vectors. The method described here uses homologous recombination machinery of *Escherichia coli* BJ5183 to construct plasmids used in generation of adenoviral vectors. With this method, no ligation steps are involved in generating the plasmids, and any region of the adenoviral genome can be easily modified. Briefly, the full-length adenoviral genome flanked by unique restriction enzyme sites is first cloned into a bacterial plasmid. Next, the region of the viral genome to be modified is subcloned into a bacterial shuttle plasmid, and the desired changes are introduced by molecular biology techniques. The modified viral DNA fragment is gel-purified and cotransformed with the full-length plasmid, linearized in the targeted region, into BJ5183 cells. Homologous recombination in *E. coli* generates plasmids containing the modified adenoviral genome. Recombinant virus is generated following release of the viral DNA sequences from the plasmid backbone and transfection into a producer cell line. With this method, homogeneous recombinant adenoviruses can be obtained without plaque purification.

Key Words: Adenovirus; modification of adenoviral genome; full-length plasmid; bacterial homologous recombination system; *E. coli* BJ5183.

1. Introduction

Recombinant adenoviruses (Ads) are one of the most commonly used viral vectors for a variety of therapeutic and research purposes both in vitro and in vivo. Several attractive features of Ads have made them the vectors of choice for many applications in gene expression, gene therapy, and oncolytic vectors. These attractive features include the following: the virus is able to infect a

From: *Methods in Molecular Medicine, Vol. 130:*
Adenovirus Methods and Protocols, Second Edition, vol. 1:
Adenoviruses Ad Vectors, Quantitation, and Animal Models
Edited by: W. S. M. Wold and A. E. Tollefson © Humana Press Inc., Totowa, NJ

broad spectrum of both dividing and nondividing cell types; the viral genome is well studied, has high stability, and is easily manipulated; and the virus can be grown and purified to high titers under Good Manufacturing Practice conditions. Methods to construct adenoviral vectors are well documented, and they vary immensely in their approach, efficiency, and starting materials. Several published protocols and commercially available kits allow for construction of recombinant Ads in which a trangene expression cassette replaces the E1 or E3 region. However, manipulation of other portions of the genome is often very difficult because of the large size of Ad genome plasmids. Most alterations that have been explored for easier or more efficient Ad vector construction have not addressed this issue.

The primary goals for new vector generation include: (1) accuracy, so that the new vectors are of the desired structure; (2) ease of use entailing general laboratory techniques and reagents; and (3) efficiency in both time and labor. The method described here meets these criteria. This method utilizes plasmids that contain all or portions of the Ad genome; smaller plasmids are useful for manipulations, whereas the larger genomic plasmids provide the remaining regions. In this case, recombination takes place in a strain of *Escherichia coli*, BJ5183, that is *recBC sbcBC*. This method was originally described by Chartier et al. *(1)*, who demonstrated the construction of several Ad vectors with insertions into the E1 and E3 regions. Later, He et al. *(2)* used similar methods to develop a series of plasmids to facilitate recombinant Ad construction for vectors with transgenes inserted in the E1 region. Since then, numerous Ad recombinants have been generated using these BJ5183 recombination methods. Although these papers describe how to make recombinants with alterations in the E1 or E3 region, any region is amenable for changes if the appropriate plasmids are available.

This article describes how to efficiently generate recombinant Ads using recombination in *E. coli* BJ5183 based on the references above. However, the method has been expanded to include other regions of the Ad genome and optimized so that it provides desired Ad genomic structure with nearly 100% efficiency.

This method of vector generation involves the construction of plasmids containing full-length adenoviral genomes generated in *E. coli* BJ5183. All targeted modifications of the viral genome are introduced into the full-length plasmid by homologous recombination. Recombinant vector is subsequently generated following excision of the Ad genome from the plasmid and transfection into a producer cell line, such as 293. The *E. coli* BJ5183 strain used in this system is *rec*A positive and possesses all of the necessary enzymes to execute the recombination event between overlapping DNA sequences present between the shuttle vector and the full-length Ad genomic plasmid. This procedure is

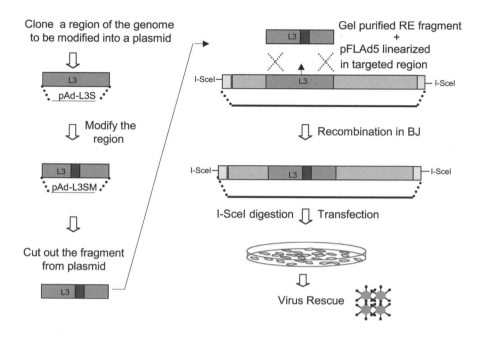

Fig. 1. Steps involved in generation of recombinant adenoviral vectors using the BJ recombination system.

graphically outlined in **Fig. 1**. The full-length adenoviral genome bordered by two unique restriction sites is first cloned into a bacterial plasmid; this plasmid, containing the full-length viral DNA, is referred to here as pFL. In a second plasmid, the viral region to be modified, derived from anywhere in the Ad genome, is cloned into a bacterial plasmid and must contain homology to sequences present in the pFL on both sides of the region to be modified; this is referred to here as a shuttle plasmid. All of the desired changes, such as deletions, insertions, or mutations, are performed in the shuttle plasmid. The modified DNA segment is cleaved from the plasmid backbone by restriction enzyme digestions followed by purifications from electrophoresis gels. The full-length plasmid (pFL) is linearized in the targeted region by digestion with an appropriate restriction enzyme, but is not gel-purified. This linearized full-length plasmid and the gel-purified modified fragment of the shuttle plasmid are cotransformed into BJ5183. Two homologous recombination events, between the free ends of the modified fragment and their homologous sequences in the linearized pFL, lead to generation of recombinant pFL containing adenoviral genomes with the expected changes. Because the yields of plasmid DNA from BJ5183 are very low, the plasmid DNA is subsequently transformed into *E. coli* DH5α cells to generate large quantities of plasmid

DNA. In the pFL example described in **Fig. 1**, two rare restriction enzyme sites, I-SceI, were introduced upstream of the left inverted terminal repeat (ITR) and downstream of the right ITR. This allows for easy and convenient excision of the full-length adenoviral genome from the plasmid prior to transfection into mammalian cells. At least one of the ITRs must be freed from the plasmid backbone prior to transfection into cells to allow for DNA replication, as exemplified by previous strategies that cut the full-length vector only once. Transfection of the closed-circular pFL (i.e., not liberated from the plasmid backbone) will not generate viruses.

Although the BJ5183 recombination method utilizes standard molecular biology techniques, several steps and details require extra attention because of the size of the plasmids (>36 kb) and special features of the BJ5183 strain. Several factors influence recombination frequency in BJ5183. First, linearization of pFL prevents the growth of background colonies containing the parental plasmid and enhances the recombination process. The percentage of positive colonies increases (approx 100%) when pFL is linearized precisely in the recombination-targeted region. Second, the length of homology on either end of the targeted region affects the efficiency of recombination. Although successful recombination can be achieved with as little as 200 bp of homology on either side of the targeted region, the frequency of recombination is significantly lower. Increasing the length of the homology region to more than 500 bp significantly enhances the frequency of recombination. Third, the stability and total size of the full-length plasmid are sensitive to the manipulations. Plasmids in BJ5183 cells tend to rearrange when the total size exceeds 40 kb, so it is essential to keep the size of the plasmid below 40 kb. Fourth, the growth conditions can profoundly affect the efficiency of pFL vector generation. Growth conditions such as incubation temperature (30°C vs 37°C) and the amount of ampicillin (50 vs 100 µg/mL) in the culture medium can influence the growth of *E. coli* BJ5183 and DH5α cells harboring large plasmids. Finally, in some instances the type of transgene and orientation of the expression cassette present in the adenoviral genome influence the stability of the full-length plasmid.

A number of different Ad vectors, including replicating and nonreplicating, gutless, and transgene-expressing vectors, have been made using the optimized procedures described here. Although this method may tolerate some changes, most modifications will compromise the process. Various shuttle and full-length plasmids can be constructed to tailor this method for modification of various regions of the Ad genome.

2. Materials

2.1. Plasmids, E. coli *BJ5183, and Reagents Necessary for Recombination*

1. Plasmids with small regions of the Ad genome that are targets for manipulation.
2. Large genomic plasmid containing most or all of the Ad genome for recombination.
3. *E. coli* strains BJ5183 and DH5α.
4. Media: Luria-Bertani (LB) broth and LB agar plates plus 100 µg/mL ampicillin, LB broth plus 100 µg/mL ampicilllin; SOC medium (2%) tryptone, 0.5% yeast extract, 10 mM NaCl, 2.5 mM KCl, 10 mM MgCl$_2$, 10 mM MgSO$_4$, 20 mM gluclose.
5. Incubators and shakers at 30°C and 37°C.
6. Restriction enzymes and buffers, Qiagene plasmid purification kit (cat. no. 19046).
7. Buffers: RF1: 100 mM RbCl (12 g/L), 50 mM MnCl$_2$ (MnCl$_2$·2H$_2$O; 9.9 g/L), 30 mM potassium acetate (2.94 g/L), 10 mM CaCl$_2$ (CaCl$_2$·2H$_2$O; 1.5 g/L), 15% (w/v) glycerol (150 g/L); adjust pH to 5.8 with 2 M acetic acid. Filter-sterilize, store at 4°C.
8. RF2: 10 mM MOPS (2.08 g/L), 10 mM RbCl (1.2 g/L), 75 mM CaCl$_2$ (CaCl$_2$·2H$_2$O; 11.25 g/L), 75% (w/v) glycerol (150 g/L); adjust pH to 6.8 with NaOH. Filter-sterilize, store at 4°C.

2.2. Transfection

1. Lipofectamine.
2. Suitable cells, such as PER.C6, and appropriate medium.

3. Methods

3.1. Preparation of Frozen Competent E. coli *BJ5183*

1. Streak BJ5183 cells from a frozen stock onto LB agar (without any antibiotic); plate and incubate overnight at 37°C.
2. Pick a single colony and inoculate 2 mL of LB broth (no antibiotic); incubate overnight at 37°C in a shaker incubator (200 rpm).
3. Add 1 mL of the overnight culture to 500 mL of LB broth (no antibiotic) and incubate at 37°C with shaking (200 rpm) until the optical density (OD) reaches 0.5 at 550 nm (usually takes 3–5 h).
4. Transfer the culture to a sterile centrifuge bottle and chill on ice for 10 min.
5. Pellet the cells by centrifugation at 3000g (4°C) for 10 min.
6. Pour off the supernatant and remove the residual medium by aspiration with pipet.
7. Gently resuspend the cell pellet in 150 mL RF1 (one-third volume of the original culture volume) and incubate the cell suspension on ice for 30 min.
8. Pellet the cells by centrifugation at 3000g for 10 min at 4°C.
9. Decant the supernatant thoroughly and gently resuspend the cells in 10 mL of RF2; leave the cell suspension on ice for 10 min. Meanwhile, chill 1.5 mL Eppendorf centrifuge tubes (up to 100 tubes) on ice and place Eppendorf tube boxes at –80°C.
10. Distribute 100 µL of cell suspension into each chilled tube, transfer tubes into chilled boxes, and store at –80°C. The cells remain competent and viable for at least 1 yr.

3.2. Preparation of Plasmid DNA

1. Preparation of pFL: cut 5 µg of the plasmid DNA containing the full-length Ad genome with an enzyme that will linearize the plasmid. Ideally, the restriction site will be within the region where recombination should occur. Extract the DNA once with an equal volume of phenol/chloroform/isoamyl alcohol, and once with chloroform. Precipitate the DNA by the addition of 2 vol of 100% ethanol, wash pellet once with 70% ethanol, allow to air-dry, and then resuspend the DNA in 25 µL of sterile water.
2. Preparation of the shuttle plasmid: at the same time, excise the insert from the shuttle plasmid, starting with approx 10 µg DNA to obtain purified insert DNA. Gel-purify fragment by glass wool technique. Resuspend in 10–15 µL of sterile water.

3.3. Transformation of Competent E. coli BJ5183 Cells

1. Thaw competent BJ5183 cells on ice for 5 min.
2. Mix DNA samples using approx 100 ng of linearized pFL and 500 ng of gel-purified fragment for a total volume not exceeding 10 µL in an Eppendorf tube and add it to one tube of competent cells (100 µL); tap gently and return the tube to ice. Also set up a control reaction consisting of the linearized full-length plasmid DNA only.
3. Heat-shock by incubating at 42°C for 60 s and then immediately transfer the tube back to ice for 5 min.
4. Add 1 mL of SOC and shake the tube at 30°C for 20–30 min.
5. Pellet the cells by centrifugation at 1500g for 1 min and pour off about 800 µL of medium; resuspend the cell pellet in the remaining medium by gently vortexing the tube. Spread the cells over two LB amp (100 µg amp/mL) plates (preferably freshly made plates).
6. Incubate the plates at 30°C for 24–36 h.

3.4. Extraction and Analysis of Mini-Prep DNA From E. coli BJ5183 Cells

1. Compare number of colonies to control plate and select very small colonies (*see* **Note 1**). Pick six BJ5183 colonies and grow each in 2 mL LB amp (amp 100 µg/mL) broth at 30°C in a shaker incubator (200 rpm; *see* **Note 1**).
2. Transfer 1.5 mL of the culture to 1.5-mL Eppendorf centrifuge tubes and pellet the cells by centrifugation at 10,000g for 30 s.
3. Pour off the medium and suspend the cell pellet in 100 µL of P1 solution by vortexing (Qiagen maxi-prep kit–plasmid buffer set; cat. no. 19046).
4. Add 100 µL of P2 solution; mix the contents by inverting the tube two to three times. *Do not vortex.*
5. Add 100 µL of P3 solution and mix the contents gently by inverting the tubes two to three times. *Do not vortex.*
6. Centrifuge immediately at 15,000g for 7 min at room temp.
7. Collect the supernatant into a fresh Eppendorf tube and add 500 µL of absolute ethanol and mix the contents by inverting the tube several times.

8. Immediately pellet the DNA by centrifugation at 15,000g for 7 min at room temp. Pour off the supernatant, spin briefly, and remove residual ethanol using a pipet tip.

9. Air-dry the DNA pellet (or 1 min in a Speedvac® concentrator) and suspend in 20 µL sterile water.

10. Digest 10 µL of DNA with a suitable restriction enzyme, and analyze the DNA by agarose gel electrophoresis and ethidium bromide staining. Select three positive clones that show the expected restriction enzyme digestion pattern.

11. The remaining half of the DNA from the selected clones is used to transform DH5α cells (for subcloning efficiency) by heat-shock method as described in **Subheading 3.3.** (*see* **Notes 2** and **3**).

12. Plate the transformation mix on LB amp (50–100 µg/mL amp) plates and incubate at 37°C.

13. Usually less than 50 DH5α colonies are obtained. More than 50 colonies normally suggests a high background, and the plate may not yield the proper clones. Pick a single colony and inoculate it into 2 mL of LB amp (50 µg/mL amp) broth (*see* **Note 2**). Incubate at 37°C for 8–16 h. Isolate plasmid DNA using standard mini-prep procedures. DNA analyses can be performed by restriction enzyme digestion and agarose gel electrophoresis to confirm the correct structure.

14. If a larger quantity of DNA is desired, it is necessary to streak the mini-prep culture onto a LB amp (50 µg/mL of amp) plate. The following day, use a single colony to inoculate 200 mL of LB (containing 50 µg/mL of amp). Commercially available maxi-prep kits such as Qiagen can be used.

3.5. Transfection of Full-Length Plasmid DNA Into Complementing Cells Using Lipofectamine

1. In a 25-cm^2 tissue culture flask, seed 7×10^6 complementing cells such as 293 or PER.C6 cells in 5 mL of Dulbecco's modified Eagle's medium (DMEM) supplemented with 10% fetal bovine serum (FBS) and incubate at 37°C in a CO_2 incubator for 24 h.

2. The following day, digest 4 µg of pFL DNA with I-SceI or suitable enzyme to liberate viral DNA sequences from the plasmid backbone. Digestion is typically carried out in 30 µL reaction volume at 37°C for 2 h.

3. Increase the volume of digestion mix to 500 µL by addition of 470 µL of sterile water and extract the digested DNA once with equal volume of phenol-chloroform-isoamyl alcohol, and once with chloroform. Precipitate the DNA by addition of two volumes of salted ethanol (absolute ethanol containing 2% potassium acetate), wash the pellet with 70% ethanol, and suspend the DNA pellet in 20 µL sterile water.

4. Make solution A by adding 3–5 µg of the digested DNA to serum-free DMEM to give a total volume of 100 µL.

5. Make solution B by mixing 40 µL of Lipofectamine and 60 µL serum-free DMEM.

6. Add solution B to solution A (not A to B) and mix gently by tapping the tube; incubate for 30 min at room temperature.

7. Just before transfection, replace the growth medium in the flask with 5 mL of serum-free DMEM medium and further incubate the flask at 37°C in CO_2 incubator for 30 min.

8. Add 2.3 mL of serum-free DMEM to the 0.2 mL DNA-Lipofectamine mixture (solutions A and B, from **step 6**) and mix by pipeting.

9. Aspirate the medium from the cells and add DNA-Lipofectamine mixture to the cells in the 25-cm^2 flask and incubate in the CO_2 incubator for 2–3 h.

10. Add 2.5 mL DMEM containing 20% FBS and 10 mM $MgCl_2$ and continue incubation in the CO_2 incubator at 37°C for 12 h.

11. Replace the transfection medium with suitable medium containing 10% FBS and incubate for 24 h at 37°C in CO_2 incubator.

12. Wash the cells with 2 mL trypsin and then add 0.5 mL trypsin; incubate for about 2 min at 37°C.

13. Add 12 mL DMEM containing 10% FBS and transfer to 75-cm^2 flask; incubate in CO_2 incubator at 37°C and monitor the cells every day for appearance of cytopathic effect (CPE).

14. At full CPE, harvest the cells into the medium; subject it to three cycles of freeze–thaw to make crude virus lysate (CVL). CVL is used as inoculum to make a large-scale preparation of the virus.

15. Blind-passage the transfected cells once if CPE is not seen within 7 d.

4. Notes

1. When selecting colonies following transformation of BJ5183, choose colonies that appear very small, as they are more likely to contain the correct plasmid. Also, compare experimental plate to control plate. If there are many colonies on the control plate, it is less likely that the experimental plate will contain colonies with the desired recombinant.

2. The yield of plasmid DNA from BJ5183 is very low, possibly because of the tendency of these cells to generate plasmid concatamers. Therefore, it is necessary to use DH5α cells for large-scale preparation of plasmid DNA.

3. The plasmid DNA derived from BJ5183 cells may be unstable in some cases; therefore, it is recommended to transform DH5α cells on the same day as DNA extraction from BJ5183 cells.

References

1. Chartier, C., Degryse, E., Gantzer, M., Dieterlé, A., Pavirani, A., and Mehtali, M. (1996) Efficient generation of recombinant adenovirus vectors by homologous recombination in *Escherichia coli*. *J. Virol.* **70**, 4805–4810.

2. He, T.-C., Zhou, S., Da Costa, L. T., Yu, J., Kinzler, K. W., and Vogelstein, B. (1998) A simplified system for generating recombinant adenoviruses. *Proc. Natl. Acad. Sci.* **95**, 2509–2514.

6

Production and Release Testing of Ovine Atadenovirus Vectors

Gerald W. Both, Fiona Cameron, Anne Collins, Linda J. Lockett, and Jan Shaw

Summary

Gene-directed enzyme prodrug therapy (GDEPT) is an emerging approach for the treatment of cancers. A variety of viral vectors have been used to deliver genes that encode the relevant enzymes, and some have been tested in clinical trials. To ensure the potency and efficacy of such vectors and to obtain regulatory approval to administer them to humans, it is necessary to develop a suite of assays that provide quality assurance. New GDEPT vectors based on ovine atadenovirus and *Escherichia coli* purine nucleoside phosphorylase (PNP) have been developed for first time use in humans in a phase I trial for the treatment of prostate cancer. Here we describe methods for their production together with several quality-control assays. In particular, a functional cell killing assay was devised to measure the potency of PNP-GDEPT vectors, the principles of which could easily be adapted to other systems.

Key Words: Ovine atadenovirus; vectors; PNP-GDEPT; prostate cancer.

1. Introduction

Human adenovirus vectors have been widely used in the laboratory and clinic, and procedures for their construction and manufacture are well known *(1)*. However, attention is now also being directed to nonhuman adenovirus vectors partly because they are not neutralized by immunity that exists in humans as a result of natural infection, which may provide an advantage for gene delivery in vivo. The nonhuman vector derived from an ovine adenovirus isolate OAdV287 (OAdV) has been developed as a gene delivery vector *(2)*. OAdV is the prototype of a new genus known as the atadenoviruses *(3)*. These viruses have very different biological properties compared with mastadeno-viruses, the best studied group. Some clues as to the function of certain genes

From: *Methods in Molecular Medicine, Vol. 130:*
Adenovirus Methods and Protocols, Second Edition, vol. 1:
Adenoviruses Ad Vectors, Quantitation, and Animal Models
Edited by: W. S. M. Wold and A. E. Tollefson © Humana Press Inc., Totowa, NJ

have been obtained, but there are numerous genes with no known homologs whose functions are still being elucidated *(4)*.

Vectors OAdV220 and OAdV623 express the *Escherichia coli* PNP gene from different promoters (*see* **Subheading 2.1.**, **item 3**). These were constructed and tested in preclinical GDEPT studies for prostate cancer *(5–8)*. Procedures for the construction of plasmids and rescue of infectious OAdV vectors from transfected DNA have been described *(9,10)*. Because OAdV replication is abortive in nonovine and human cell lines, these vectors were propagated in fetal ovine skin (HVO156) or lung (CSL503) cells *(11)*, which are established, adherent cell lines that have not been immortalized. In this work we have used CSL503 cells because master and working cell banks have been established and tested for adventitious agents.

OAdV623 and OAdV220 lysed the production cells very efficiently, necessitating changes to earlier methods of harvest. Here we describe the modified manufacturing procedure together with quality-control assays that were used routinely. Additional assays were used by a contract manufacturer to characterize vector suitable for a phase I clinical trial.

2. Materials

2.1. Cell Lines and Viruses

1. CSL503 fetal ovine lung cells (CSL Ltd, Parkville, VIC) were frozen as a cell bank at doubling 41. Cells were grown in minimum essential medium (MEM) supplemented with nonessential amino acids (0.1 mM), glutamine (2 mM), N-2-hydroxyethylpiperazine-N-2-ethanesulfonic acid (HEPES) (28 mM) (all from Gibco; Invitrogen Corp, Carlsbad, CA), penicillin/streptomycin (50 IU/50 µg/L) (Gibco/BRL), and fetal bovine serum (FBS) (10%) (Thermo Electron Corp, Melbourne, Victoria, Australia).

2. PC3 human prostate cancer cells (ATCC CRL-1435) for testing virus potency were frozen down as a working cell bank (1–2.5 × 10^6 cells/vial) at doubling 18 and used within five passages. For cell growth, RPMI with glutamine and without NaHCO$_3$ (Gibco/BRL), made according to manufacturer's instructions, was supplemented with 10% FBS and penicillin/streptomycin (50 IU/50 µg/L). Typically, cells were passaged twice per week at a split ratio of 1:5 by treating them lightly with trypsin.

3. Recombinant virus OAdV623 contains the wild-type genome with a gene cassette inserted in a nonessential site (III) between the E4 and RH transcription units *(2)*. The gene cassette comprises the PSMA enhancer and rat probasin promoter (the combination provides prostate cell-specific expression) linked to the *E. coli* purine nucleoside phosphorylase coding sequence and bovine growth hormone polyadenylation signal *(8)*. OAdV220 contains a similar cassette in which the RSV promoter is used. In this case the cassette is inserted in site I between the pVIII and fiber genes *(5)*.

4. OAdV623 formulated in Tris sucrose PEG buffer containing 10 μM cationic lipid CS087 (not commercially available) is known as FP253. CS087 is a Tris-conjugated cationic lipid (trilysine-capryloyl-tris-trilaurate ditrifluoroacetate [T-shape]; empirical formula: $C_{66}H_{130}N_8O_{10} \cdot 2C_2F_3O_2$; molecular weight: 1421.81) *(12)*. Lipid formulation may enhance virus infectivity in cell types that lack the viral receptor *(13)*. FP253 is stored at –80°C in a minimum volume of 200 μL in 2-mL borosilicate vials with snap-sealed rubber stoppers.

5. A reference batch of OAdV623 was set aside in 25 μL-aliquots and used as a standard in gel analysis and assays of potency and infectivity. Compiling results for the reference batch in these assays made it possible to monitor interassay variation and the long-term performance of individual assays.

2.2. Stock Solutions and Buffers

1. Tris-ethylene diamine tetraacetic acid (EDTA) (TE): 10 mM Tris-HCl, pH 8.0, 1 mM EDTA.
2. Cell lysis buffer: 0.1% sodium dodecyl sulfate (SDS), 10 mM Tris-HCl, pH 8.0, 1 mM EDTA.
3. Virus dissociation buffer: 0.1% SDS in 10 mM Tris-HCl, pH 8.0.
4. CsCl solutions: to prepare cesium chloride solutions of density 1.5 g/cm^3, 1.35 g/cm^3, and 1.25 g/cm^3 in TE, pH 8.0, separately weigh 45.4, 35.5, or 27.6 g of CsCl, respectively, and add TE until the final weight of each solution is 100 g.
5. TSP buffer for virus storage: 10 mM Tris-HCl, pH 8.0, 8.5% (w/v) sucrose, 0.5% (v/v) PEG400. Sterilize by filtration (0.2-μm; Sartorius Minisart RC 15).
6. TS buffer: as for TSP buffer but without PEG400.
7. Fludarabine phosphate: "Fludara" (Schering Pty. Ltd., Alexandria, NSW). A 50-mg vial made up to 1 mL in deionized water is 136.9 mM. Dilute to 1 mM stock. Freeze in 0.5-mL aliquots at –80°C. Discard after one thaw.
8. Dulbecco's phosphate-buffered saline (DPBS): add 10 mL of 100X CaCl$_2$ and 10 mL of 100X MgCl$_2$ to 980 mL of sterile PBS (Oxoid). 100X stocks are CaCl$_2 \cdot$2H$_2$O (1.33 g) or MgCl$_2 \cdot$6H$_2$O (1.0 g) dissolved in and separately made up to 100 mL with deionized water.
9. MTS: Cell Titer 96® AQ$_{ueous}$ reagent powder (Promega Corp., Madison, WI). Add 42 mg MTS powder to 21 mL of DPBS. Adjust to pH 6.0–6.5 with 1 M HCl if necessary. Filter-sterilize (0.2 μm; Sartorius). Protect from light. Store at –20°C. Stable for up to 6 mo.
10. 0.92 mg/mL PMS in DPBS. Filter and store as for MTS. Stable for up to 6 mo.
11. Trypsin (0.25%) /EDTA (Sigma).
12. Gel running buffer: 24 mM Trizma base, 190 mM glycine, pH 8.3, 1% SDS.
13. Gel loading buffer: 50 mM Tris-HCl, pH 6.8, 10% SDS, 0.1 M dithiothreitol (DTT), 10% glycerol, 0.1% bromophenol blue.

2.3. Safety Notes

All work with OAdV623 must be carried out in a class II biohazard cabinet in a PC2 (or international equivalent) containment laboratory. Virus-containing material should be autoclaved before disposal. Laboratory coats, safety glasses, and nitrile gloves should be worn. Fludarabine phosphate is a cytotoxic drug and should be handled according to the material safety data sheet provided.

3. Methods

3.1. Propagation of CSL503 Cells

Seed CSL503 cells at approx 15,000 cells/cm^2 in 150-mm tissue culture dishes or 850-cm^2 roller bottles (Corning). Propagate them in MEM plus supplements and passage one in three once a week by treatment with 0.25% trypsin/EDTA. Refeed the cells mid-week. Infect with virus when the cells reach approx 90% confluence (density ~35,000–50,000 cells/cm^2, 3–4 d after the last passage). Cells can be passaged to doubling 54 with no major effect on virus yields.

3.2. Infection

Retrieve OAdV infectious stock from the –80°C freezer and thaw on ice. Use a multiplicity of infection of approx 1 TCID$_{50}$ per cell (for TCID$_{50}$ units, *see* **Subheading 3.6.7.**) assuming there are approx 5×10^4 cells/cm^2 present at infection. Remove old-growth medium from adherent CSL503 cells and add the infectious virus directly to fresh MEM plus supplements (15 or 85 mL per 140-mm dish or 850-cm^2 roller bottle, respectively, which is half of the volume used for cell propagation). Place dishes of infected cells on a rocking platform at its slowest setting in a CO$_2$ incubator at 37°C. Roller bottles are gassed with filtered (0.2-μm, Sartorius) 5% CO$_2$ in air, sealed, and placed on a roller rack at 1 rpm in a 37°C room. A cytopathic effect (CPE) on the cells can be observed under the microscope and should be complete in 3–4 d.

3.3. Virus Harvesting and Concentration

This protocol is adapted from a procedure used to isolate human AdV5 from culture medium *(14)*. Except for centrifugation steps, where solutions containing virus are in sealed containers, all work is carried out in a class II biohazard cabinet.

1. After CPE has developed, medium is transferred to conical-bottomed centrifuge tubes (Nalgene 175-mL or Corning 50-mL, depending on the scale of the virus preparation) and floating, intact cells are removed by centrifugation at 1300g for 10 min. Retain the supernatant and hold it at room temperature. Any cell pellet may be combined with other residual cells (*see* **Note 1**).
2. Measure the total volume of cell culture medium in the conical bottom centrifuge tubes. Weigh out (NH$_4$)$_2$SO$_4$ (242.3 g for each L of medium), add it to the liquid, and mix thoroughly to achieve a saturation of 40%. Incubate the mixture at room

temperature for at least 2 h with occasional gentle mixing by inversion to allow the precipitate to form (*see* **Note 2**).

3. Centrifuge to recover the pellet containing virus (1600g for 15 min). Pour off the supernatant and gently dissolve the pellet in a minimal volume of PBS. Because the supernatant may contain some residual virus, treat with 0.2–2% bleach for 30 min at room temperature before discarding.

3.4. Purification of OAdV by Ultracentrifugation

1. The virus is first fractionated by velocity centrifugation on CsCl step gradients. Each step gradient can accommodate 4.5 mL of crude virus concentrate. Add 2.5 mL of 1.25 g/cm^3 CsCl to a Beckman Ultraclear centrifge tube (14 × 89 mm for the SW41 rotor), then underlay it with 2.5 mL of 1.35 g/cm^3 and 2 mL of 1.5 g/cm^3 solution using a 5-mL pipet with the Pipet Aid set on "slow." Carefully layer the crude virus concentrate on top using a sterile Pasteur pipet. Centrifuge in a Beckman ultracentrifuge using an SW41 rotor at 35,000 rpm (151,000g) for 2 h at 4°C. Balance opposing buckets to within 5 mg.

2. The viral band (the lowest band in the gradient, usually about halfway down) is best viewed against a black background. Mark its position on the tube and carefully remove the overlying solution with a sterile Pasteur pipet. Recover the viral band in a minimal amount of CsCl solution.

3. Pool virus samples from all tubes, gently mix, and place half the solution in each of two clear centrifuge tubes. Fill these with 1.35 g/cm^3 CsCl in TE and centrifuge to equilibrium in the SW41 rotor for 20 h at 35,000 rpm (151,000g) at 4°C. The virus band will be found in the top one-third of the gradient (**Fig. 1**). Recover the virus as described above. For greater purity, repeat the centrifugation step using one gradient only.

3.5. Desalting Concentrated Virus

1. Place a Nap25 column (Amersham/Pharmacia) in a retort stand (wiped with 70% ethanol). Equilibrate the column by 3 × 5-mL applications of virus storage buffer item.

2. Mix and apply the viral band from the final gradient (<2 mL) to the equilibrated column. Allow the sample to run into the column; then apply 10 × 0.5-mL of virus storage buffer and collect the eluted fractions in sterile tubes.

3. Identify fractions containing the most virus using a PicoGreen binding assay (Molecular Probes, Eugene, OR) as follows:
   a. Add 5 µL of each fraction to 100 µL of virus dissociation buffer followed by 100 µL of PicoGreen diluted 1:400 in deionized water. Include a blank with no virus added.
   b. Mix well and add a 100 µL aliquot of each sample to individual wells of a black 96-well plate. Read the fluorescence at 485$_{nm}$ excitation/535$_{nm}$ emission in a fluorescent plate reader (Wallac Isoplate Victor^2; Perkin Elmer).
   c. Pool the three to four fractions with the highest PicoGreen readings, and mix gently but completely.

Fig. 1. Bands of OAdV623 after centrifugation to equilibrium on CsCl gradients.

4. Gently take up the virus in a 2.5-mL syringe using a 38-mm blunt-ended needle (18-gage; Terumo). Attach a filter unit (Millex GV 0.22-µm × 13-mm; Millipore) to the syringe, depress the plunger, and collect the filtered virus in sterile Eppendorf tubes. Remove 6 × 20-µL and 1 × 100-µL samples for quality control analysis; dispense the remainder of the virus in 2- to 300-µL aliquots and store at –80°C.

5. If desired, OAdV623 may be formulated with cationic lipid to facilitate infection of cells that have a low number of viral receptors.
   a. Calculate the total volume of formulated virus required. In a polypropylene tube, dilute purified virus at 4°C with TSP to twice the final concentration desired, allowing for 5–10% overage.
   b. In a glass vial, dilute CS087 in TSP to 20 µ*M* (twice the final concentration required) in half the final volume desired. Allow it to warm to room temperature.
   c. Add one-half volume of OAdV623 in TSP to the glass vial and mix gently but thoroughly. Gently agitate the suspension at 40 rpm for 60–90 min at room temperature.
   d. Store the formulated virus in 0.2-mL aliquots in 2-mL glass vials at –80°C.

3.6. Quality Control Assays

Below we summarize assays for the vector that are routinely used to verify identity, purity, and genetic stability. We also describe a new assay designed to measure vector potency. The latter is especially important because OAdV220 and OAdV623 carry the components of a cell killing system intended for the treatment of solid tumors.

3.6.1. pH of the Virus

Measure the pH of the virus preparation before freezing by placing a 20-µL droplet on the window of a precalibrated micro-pH meter (IFSET KS723,

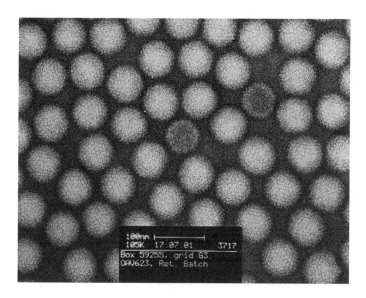

Fig. 2. Analysis of OAdV623 by electron microscopy after negative staining. The bar represents 100 nm.

Shindengen Electric Mfg. Co. Ltd, Tokyo). It should be pH 8.0, but values between pH 6.5 and 8.3 have been obtained where the virus had satisfactory infectivity and potency.

3.6.2. Electron Microscopy

Monitor virus purity and integrity by negative staining and transmission electron microscopy. Desalt the 20 μL sample (to remove sucrose) by spinning it briefly (750*g* for 2 min) through a Microspin G-25 Column (Amersham/ Pharmacia) in an Eppendorf tube. Keep the purified sample on ice and have it processed as soon as possible by the electron microscopist. A typical OAdV preparation is shown in **Fig. 2**.

3.6.3. Virus Particle Number

The concentration of virus particles (vp) in a preparation can be determined as previously described *(15)*.

1. Mix a 100-μL aliquot of the virus with an equal volume of virus dissociation buffer in an Eppendorf tube and heat at 65°C for 20 min.
2. Read the optical density (1-cm path length) at wavelengths of 310, 280, and 260 nm using a spectrophotometer (Shimadzu UV-1601) with buffer alone as blank. The OD_{310nm} reading should be <0.02. The OD_{260nm} reading should be in the range of >0.1 to <0.8. If >0.8, dilute the sample and read it again (*see* **Note 3**). The $OD_{260/280nm}$ ratio is usually between 1.19 and 1.25.

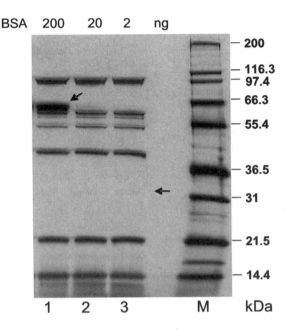

Fig. 3. Analysis of OAdV623 virus by SDS-PAGE. CsCl-purified virus (4×10^9 vp/ lane) was analyzed on a 4–20% gradient gel. The horizontal arrow indicates the position of 32K protein, which stains poorly with silver. To obtain an estimate of virus purity, samples were spiked with known amounts of bovine serum albumin (BSA) as shown.

3. The concentration of virus particles is calculated from the OD_{260nm} reading, assuming 1.3×10^{12} vp per OD_{260nm} unit. This figure is based on the extinction coefficient determined for AdV5 *(16)* and takes into account the smaller genome size of OAdV (35,937 bp vs 29,575 bp, respectively).

3.6.4. Analysis of Viral Proteins

Virus purity is monitored by electrophoresis in SDS-polyacrylamide gels. CSL503 cell lysate diluted 1:16 to 1:64 may be used as a control for contamination by cellular proteins.

1. Heat the lysate, virus test, and reference samples (~4×10^9 vp per lane) and protein molecular-weight standards (e.g., Invitrogen Mark 12™) at 98–100°C in an equal volume (15–25 µL) of gel loading buffer.
2. Cool and load the samples onto a 4–20% preformed gradient gel (Gradipore, French's Forest, NSW). Alternatively, an 8–18% gradient gel can be used.
3. Electrophorese samples in gel running buffer at 100 V until the indicator dye leaves the bottom of the gel and then for a further 20 min.
4. Viral proteins (**Fig. 3**) are detected by silver staining (SilverXpress, Invitrogen, Carslbad, CA) according to the instructions supplied with the kit.

3.6.5. Analysis by Restriction Enzyme Digestion

1. The nucleotide sequence of OAdV is known (Genbank Accession number U40839). As long as the nucleotide sequence of the inserted gene is also known, restriction digestion can be used to confirm the presence of that gene and the integrity of the OAdV genome. For OAdV623, *Bam*HI and *Eco*RV are typically used.
2. Approximately 200–300 ng of viral DNA (from ~2 × 10^{10} vp) is required per digest. This is extracted by incubating the virus with an equal volume of pronase (1 mg/mL) in TE plus 1% SDS for 2 h at 37°C.
3. Samples are then extracted twice with phenol/chloroform (saturated with TE, pH 8.0) and once with ether.
4. The DNA is recovered by precipitation with sodium acetate pH 4.8 (1/10 volume) and ethanol (two volumes).
5. Samples are frozen on dry ice for 10 min and then spun at 15,000 rpm for 10 min to precipitate DNA.
6. Pellets are washed with 70% ethanol, spun at 15,000 rpm for 5 min, air-dried, and resuspended in TE, pH 8.0.
7. After digestion, the DNA fragments are resolved by electrophoresis on a 1% agarose gel and stained with ethidium bromide. The pattern of bands is compared to the theoretical sequence and/or a reference sample digested under the same conditions.

3.6.6. Identity by Polymerase Chain Reaction

This assay is designed to identify the full-length inserted gene cassette and to confirm the identity of OAdV623 virus particles. However, by choosing suitable primers, the assay can be adapted for any adenovirus vector. The test will also confirm that the gene cassette is free from any deletions and is in the correct position within the viral genome. The assay utilizes three primer pairs, two of which (primers 1 and 2, and 3 and 4) are able to amplify the ends of the gene cassette together with their respective viral flanking sequence, while the third pair (primers 5 and 6) targets a site remote from where the gene cassette has been inserted (in the DBP gene) and acts as a positive control for any false negatives in the polymerase chain reaction (PCR). By using only the two outer primers (1 and 4), which target the viral flanking sequence, the full-length gene cassette is amplified. Because smaller fragments are preferentially amplified in a PCR with any given set of primers, any deletion from within this gene cassette will be observed as products of smaller than expected size. The primers used for OAdV623 are shown in **Table 1**.

1. For PCR analysis, purified DNA or supernatants from infected cultures can be used. Virus in the latter is diluted 1:50 with water and boiled for 5 min in a 1.5-mL Eppendorf tube additionally sealed with parafilm tape.
2. Reactions are set up using Promega 2X Master mix, appropriate primer pairs (0.2 μ*M* each), template DNA (1 μL), and Taq polymerase according to standard procedures.

Table 1
Description of Primers Used in the PCR Identity Assay for OAdV623

Primer number	Primer name	Sequence (5' to 3')
1	5' flanking primer	AACCCATTGCGTTCCTCTAAGA
2	PSME	AACTTCCCTTTCCACTTCAACTACAT
3	BGH	CGCACTGGAATCCGTTCTG
4	3' flanking primer	TGGCCTGTATGTAATGCAGTTGT
4.1[a]	3' flanking primer	GGCACCTTCCAGGGTCAAG
5	5' DNA-binding protein	TCCGACTTTAGCTTTCGGAA
6	3' DNA-binding protein	CTATGGCTACTGTAGGAGGTAGAAT

[a]Primer 4.1 is an alternative to primer 4. It appears to have greater specificity and produces a product of 2592 bp.

PCR, polymerase chain reaction; PSME, pseudomembrane; BGH, bioassayable growth hormone.

3. Cycles of 94°C for 2 min ($\times 1$), 94°C for 30 s, 60°C for 30 s, and 68°C for 3 min ($\times 40$) were used.
4. The PCR product generated by primers 1 and 4 was 2857 bp in size (**Fig. 4**). The PCR products from primer pairs 1 and 2, and 3 and 4, were 263 and 399 bp in size, respectively (**Fig. 4**). The absence of either or both of these bands indicates possible rearrangement or deletion around the cloning site. The DBP product was 428 bp in size.

3.6.7. Infectivity

The infectious titer of virus test and reference samples is determined by a limiting dilution assay. Inclusion of the reference sample allows unexpected variability in the assay to be monitored. The assay uses CSL503 cells that are permissive for OAdV replication. These cells should not be passaged for more than 54 doublings.

1. Seed the cells (10^4/well) in all wells of two 96-well plates. This provides enough wells for duplicate assays of seven $\log_{10}$ dilutions in replicates of 12 with one row of uninfected cells.
2. After overnight incubation of cells at 37°C in 5% CO_2, serially dilute the virus sample (1 in 10^4 to 1 in 10^{10}) in MEM and supplements plus FBS.
3. Remove old medium from the cells and replace it with medium containing diluted virus. Incubate the cells for a week to obtain an interim result.
4. Inspect the wells under the microscope for development of CPE and score them as positive or negative. Incubate the cells for a further week to allow the full development of CPE, particularly in the highly diluted samples.
5. Calculate the $TCID_{50}$/mL of test samples using methods originally described by Spearman and Karber *(17,18)*. An example is provided in **Table 2**.

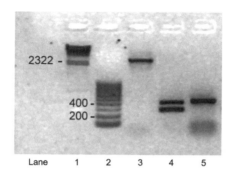

Fig. 4. Amplification of denatured OAdV623 virus DNA using the primers listed in Table 1. Lanes: (1) Lamda/HindIII DNA markers; (2) 100 bp ladder DNA markers (100–1000 bp); (3) full gene cassette (expected size 2857 bp); (4) co-amplification of the gene cassette/insertion site junctions (expected sizes 263 and 399 bp); (5) DNA-binding protein (+) control amplification (expected size 428 bp).

Table 2
Example of Analytical Test Sheet for TCID$_{50}$ Determination

	1	2	3	4	5	6	7	8	9	10	11	12	+	–	p
A-6[a]	+	+	+	+	+	+	+	+	+	+	+	+	12	0	1
B-7	+	+	+	+	+	+	+	+	+	+	+	+	12	0	1
C-8	+	+	+	+	+	+	+	+	+	+	+	+	12	0	1
D-9	+	–	–	+	+	–	–	–	+	–	+	–	5	7	0.41
E-10	–	–	+	–	–	–	–	–	–	–	–	–	1	11	0.08
F-11	–	–	–	–	–	–	–	–	–	–	–	–	0	12	0
G-12	–	–	–	–	–	–	–	–	–	–	–	–	0	12	0

[a]Row A, 10-fold dilution factor.

Count and record the total number of CPE "+" and "–" wells for each row. Calculate the proportion positive in each row by dividing the number of positives by the total number in that row. Record the answer in the "p" column for that row. For example, in **Table 2**, row D, the proportion is equal to the positive wells in row D (5) divided by the total wells (12), which equals 0.4166. The $\log_{10}$ of the 50% end point (TCID$_{50}$) in 100 µL inoculum is given by $X_0 - (d/2) + d\,(\Sigma p_i)$, where X_0 is the log of the reciprocal of the last dilution at which all wells are positive and $d = \log_{10}$ of the dilution factor (i.e., the difference between the dilution intervals. In this case $d = 1$), and p_i are the proportions, p, that were calculated above, starting with the p-value corresponding to X_0 and adding p-values from higher dilutions. Thus, from **Table 2**, $X_0 - (d/2) + d\,(\Sigma p_i) = 8 - 1/2 + 1 + 0.4166 + 0.0833 = 9.0$, but because

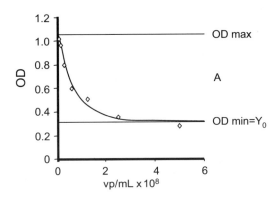

Fig. 5. Example of a first-order exponential decay curve and an analysis of a particular batch of OAdV623.

0.1 mL of inoculum was used per well, log $1/0.1 = \log 10 = 1$ must be added to this figure. Taking the antilog of the total, the 50% end point ($TCID_{50}$) titer is therefore 1×10^{10} $TCID_{50}$/mL. The vp/$TCID_{50}$ ratio can now be calculated. This is usually less than 40:1 and often less than 20:1 for individual virus preparations.

3.6.8. Functional Potency Assay

The potency of OAdV623 batches is measured in human prostate cancer PC3 cells using a cell killing assay. This quantifies cell death attributable to the combined outcome of infectivity, gene expression, and PNP activity in the presence of a pro-drug substrate, fludarabine phosphate. Cell viability is measured after 4 d by an MTS cytotoxicity assay. The cell-killing assay has been used for OAdV220, OAdV623, and FP253, but its principles can be adapted to other gene-directed enzyme pro-drug systems by using different vectors and cell lines. The outcome is expressed as the concentration of virus that corresponds to the "half-life" of the first-order exponential decay curve (VP_{50}) (**Fig. 5**). The figure provides a measure of potency that allows different virus batches to be compared while a reference batch provides a measure of consistency across assays. The result should be consistent with results from other assays in the suite of tests.

3.6.8.1. Setting Up the Test Plates

One cell culture plate setup as described here is sufficient to test two virus samples in triplicate. Each sample is tested on two separate plates with a cell standard curve and a virus reference standard on each plate (*see* **Note 4**).

1. Harvest PC3 cells by gentle treatment with trypsin, resuspend them in RPMI plus supplements at 5×10^4 cells/mL, and place them in a sterile 30-mL dispensing trough.

Table 3
Cell Standard Curve Dilution Plate[a]

Row no.	Volume of cell suspension/well (μL)	Volume of medium/well (μL)	Total cells per well
A	1000	0	50,000
B	800	200	40,000
C	600	400	30,000
D	400	600	20,000
E	200	800	10,000
F	100	900	5000
G	50	950	2500
H	0	1000	0
Total			280,000

[a]Sufficient cells to set up standard curves on three test plates.

Table 4
Number of Cells/Well in the Test Plate

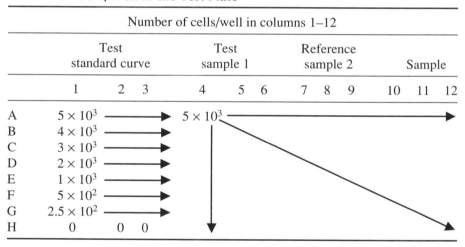

	Number of cells/well in columns 1–12											
	Test standard curve			Test sample 1			Reference sample 2			Sample		
	1	2	3	4	5	6	7	8	9	10	11	12
A	5×10^3		$\rightarrow$	5×10^3								$\rightarrow$
B	4×10^3		$\rightarrow$									
C	3×10^3		$\rightarrow$									
D	2×10^3		$\rightarrow$									
E	1×10^3		$\rightarrow$									
F	5×10^2		$\rightarrow$									
G	2.5×10^2		$\rightarrow$									
H	0	0	0									

2. Gently and thoroughly mix the cells and immediately dilute them for the cell standard curve in a plate as shown in **Table 3**.
3. Using a multichannel pipet (*see* **Note 5**), seed cells in the test plates in the following order (**Table 4**): (1) for test and reference samples, 5×10^3 cells/well (0.1 mL cell suspension) in columns 4–12 and rows A–H of all plates, seeding from top to bottom; (2) for cell standard curve, columns 1–3 and rows A–H of all plates, seed columns from left to right with 0.1 mL of diluted cells.
4. Cover and leave the plates at room temperature on a level surface for 15–30 min to allow the cells to settle evenly.

5. Wrap the edges of the plates in parafilm to prevent edge effects (such as drying out) on the cells and place them in an incubator overnight.
6. In preparation for transduction, gently aspirate the medium and replace it with 50 μL of RPMI (no serum) by adding it to the side of each well.
7. Return the cells to the incubator until the test and reference samples are ready.

3.6.8.2. SAMPLE PREPARATION

1. Thaw the test and reference samples on wet ice and mix gently.
2. Using ice-cold TS buffer, dilute at least 10 μL of each test sample to 1×10^{10} vp/mL in a 1.5-mL polypropylene Eppendorf tube (*see* **Note 6**).
3. Using polypropylene microtiter plates with 250 μL wells (Greiner Bio-one), set up twofold serial dilutions of the samples in a dilution plate used for this purpose only. To 198 μL of serum-free medium in row A, columns 4–6 and 7–9, add 22 μL of each test sample at 1×10^{10} vp/mL according to the layout shown in **Table 4**.
4. Mix and dispense samples gently into the dilution medium using the multichannel pipet with the tip immersed.
5. Similarly, set up a twofold dilution series of the reference standard in row A, columns 10–12. Following dilution, the highest virus concentration in row A is 1×10^9 vp/mL. Using a multichannel pipet, carry out doubling dilutions down the plate for columns 4–12 by transferring a volume of 110 μL into the next row. At each transfer gently mix virus with the pipet twice. Discard the diluted virus taken from row G, thus leaving row H as a control without virus.
6. For the cell standard curve wells (columns 1–3), add 22 μL TSP to 198 μL serum-free medium in row A of the dilution plate. Make twofold dilutions down the plate to row G (medium only in row H). Transfer all samples to the cell transduction plate without delay.

3.6.8.3. TRANSDUCTION

1. Carefully remove the medium from the PC3 cells by gentle aspiration.
2. Dispense 100 μL from each well of the virus dilution plate into the corresponding well of the test plate using a multichannel pipet.
3. Add samples from the most dilute wells (row H) first, gently mixing up and down once at each row before transferring. The same set of tips should be used for the whole plate to reduce virus loss through adsorption to the tips.
4. Place infected cells in the incubator for 4 h.

3.6.8.4. ADDITION OF TRANSGENE SUBSTRATE

1. For each plate, prewarm 10.1 mL of RPMI plus 20% FBS and add a 100th volume of 1 m*M* fludarabine.
2. To the infected cells, add 100 μL of this medium containing pro-drug to all wells to give a final concentration of 5 μ*M* (*see* **Note 7**).
3. Return the cells to the incubator for 4 d.

Table 5
Data Sets Generated by Microsoft Excel and Origin 7
From Analyses of Two Batches of OAdV623 and a Reference Batch

Origin parameters	Cell standard curve	OAdV623 sample 1	OAdV623 sample 2	Reference standard
R^2	1.00 (linear fit)	0.99	0.97	0.97
Y^0	—	0.26496	0.35133	0.34521
X^0	—	0	0	0
A	—	0.78621	0.75718	0.61077
t	—	2.08×10^8	2.14×10^8	1.68×10^8
Result				
ODmax	—	1.05	1.11	0.96
VP_{50}	—	1.44×10^8	1.4×10^8	1.17×10^8

3.6.8.5. Cell Viability Assay

Determine cell viability with an MTS assay (Promega Corp, Madison WI) carried out on day 4 according to the manufacturer's instructions.

1. For each test plate, mix MTS (2.2 mL) and PMS (110 µL) and prewarm RPMI plus 10% FBS (8.8 mL).
2. Gently aspirate the medium containing pro-drug from the cells, and replace it with 100 µL of medium containing MTS/PMS.
3. Incubate for 90–120 min and determine the $OD_{492nm} - OD_{650nm}$ using a spectrophotometric plate reader (LabSystems Multiskan MS). If the highest reading is less than 0.8, save the data and incubate plates for up to 4 h, reading the plate at 30- to 45-min intervals. Ideally the highest reading should be ~1.0.
4. Export the data directly into a "Microsoft Excel" format for calculations.

3.6.8.6. Data Analysis and Acceptance Criteria

Prepare the data for analysis using a first-order exponential decay curve fit (software: Origin version 7). For the standard cell curve and each test sample: calculate the mean ($OD_{492} - OD_{650}$) values from the triplicate analyses and round them to two decimal places. Subtract the mean value for blank/background wells. Remove data points if a test well is contaminated, for example, with microorganisms. Plot the data for the cell standard curve to check that the cells did not overgrow (the curve should be essentially linear).

For the test samples, fit a first-order exponential decay curve to each dose–response curve using Origin 7 software (*see* examples in **Table 5** and **Fig. 6**).

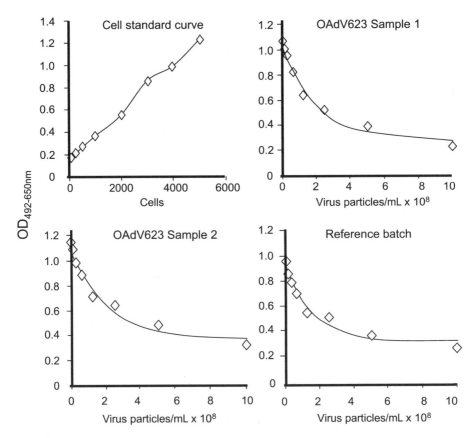

Fig. 6. Determination of the VP_{50} values for two batches of OAdV623 relative to a reference standard.

Use the equation $Y = Y_0 + Ae^{-[(X - X_0)/t]}$, where $Y = OD_{492-650}$; $X = vp/mL$; $Y_0 = Y$ offset (OD_{min}); $X_0 = X$ offset; A = amplitude; t = decay constant (*see* **Note 8**). Set $X_0 = 0$, $Y_0 \geq 0$, and $A \geq 0$. Record the parameters Y_0, A, t, and r^2 from the Origin output in the Excel spreadsheet. Calculate the ODmax ($OD_{492-650}$ reading that corresponds to no cell kill) and VP_{50} (concentration of test sample that corresponds to the "half-life" of the first order exponential decay curve) as follows:

If $r^2 \geq 0.90$, calculate VP_{50} using $VP_{50} = 0.693t$.
If $r^2 < 0.90$ and $OD_{492-650}$ of the nil virus wells is within ± 10% of $ODmax_{calc.}$, reject the data set.

Sometimes wells without virus at the edge of the plate partially dry out, but it may still be possible to salvage meaningful data. If $r^2 < 0.90$ and $OD_{492-650}$ of the nil virus wells is outside of ± 10% $ODmax_{calc.}$, remove data for nil virus wells and refit the curve using $X_0 = 0$. If $r^2 \geq 0.90$, calculate VP_{50}. If $r^2 < 0.90$, reject the data set.

3.6.9. Other Tests

There are additional tests that can be done but that are not required for routine laboratory experimentation. For virus that is destined for preclinical toxicology or clinical studies, it is necessary to determine the level of contamination by host cell proteins. Similarly, the level of contamination by cellular DNA should be determined, although in contrast to immortalized cells (e.g., 293 cells used to propagate human AdV), CSL503 cells carry no known oncogenes. Virus samples must also be tested for the presence of endotoxin, residual CsCl, and bio-burden (*see* **Note 9**).

3.7. Typical Results for Analysis of OAdV

With the experience derived from many virus preparations, the variability encountered across batches of virus has been monitored. **Table 6** shows typical results for a batch of OAdV623. For certain parameters, the range of values observed in other virus preparations is given. Importantly, values well outside of the ranges indicated were obtained on rare occasions when virus quality was thought to have been compromised, indicating that poor quality could be detected.

4. Notes

1. To maximize virus yields at harvest, any cells remaining attached to the culture vessel may be recovered by treatment with trypsin, collection in MEM (10 mL), and centrifugation. Cells are combined with residual cells from the medium and lysed with cell lysis buffer (0.25% NP40 in MEM; 1.0 mL per roller bottle) on ice for 1 h with occasional vortexing (this occurs while other manipulations are taking place). Centrifuge the combined cell lysate (1600*g* for 10 min) to pellet the cell debris. Combine this supernatant with the bulk-harvested medium. Alternatively, the cleared cell lysate can be purified separately to compare virus yields from cells vs the medium.
2. Precipitation has been allowed to continue for up to 3 d at 4°C. This can serve as a holding step if processing of the virus preparation needs to be delayed for any reason.
3. For virus samples too dilute to be measured by optical density, it is possible to obtain an estimate of the concentration by using the Picogreen assay and a virus sample of known concentration as a standard. Duplicate samples of the standard virus are diluted to approx 40 and 20 ng DNA in TE containing 1% SDS and pronase (1 mg/mL) and then heated to 56°C to dissociate the virus particles. A DNA standard curve is prepared with eight dilutions of a plasmid, in duplicate, in the range 0–50 ng/mL DNA. Test samples and standards are mixed one to one

Table 6
Typical and Acceptable Results for Preparations of OAdV623

Attribute	Typical result	Acceptable result
Appearance	Bluish, slightly viscous liquid at concentrations $<10^{12}$vp/mL: clear colorless, nonviscous liquid; at concentrations $>10^{12}$vp/mL: bluish, cloudy, slightly viscous liquid	Clear, colorless solution; free from particulates
pH	7.9^a	$6.5 - 8.3$
Identity		
1. SDS-PAGE	Conforms to reference standard	As in adjacent panel
2. Cell killing assay	Cell killing activity in range of virus titer	
3. PCR	Expected banding pattern	
Potency		
1. Infectious particle assay	6.25×10^{10} TCID$_{50}$/mL^a	Report result
2. Cell killing assay	8×10^7	$<1 \times 10^9$ vp/mL^b
3. Virus particle determination	4.7×10^{12} vp/mL from 9×850 cm^2 roller bottles^a. Range 10–30,000 vp/cell	$>2 \times 10^{11}$ vp/mL (limit of detection by OD$_{260nm}$)

Test	Actual result[a]	Specification
Purity		
1. SDS-PAGE	Conforms to reference standard and lacks additional bands	Conforms to reference standard.
2. $OD260_{nm}:280_{nm}$	1.19–1.23	1.10–1.35
3. $VP:TCID_{50}$	10–44[c]	Less than or equal to 100
Impurities/Contaminants		
1. CSL503 DNA	Not done routinely	Report result
2. CSL503 protein		Report result
3. Residual cesium chloride		Report result
Safety		
1. Endotoxins LAL assay	Not done routinely	Less than or equal to 50 EU/mL
2. Bioburden		No microbial contamination

[a]Actual result.
[b]A reference batch produced an average VP_{50} value of 1.86×10^8 with a relative standard deviation of 50.8%. Independent multiple testing by a contractor at another laboratory produced a value of 1.58×10^8 with an RSD of 30.3%.
[c]The range of values was observed for other virus batches prepared over an approx 2-yr period.
VP, a virus particle; SDS-PAGE, sodium dodecylsulfate polyacrylamide gel electrophoresis; PCR, polymerase chain reaction.

with 0.5% (v/v) Picogreen reagent and incubated without light. The fluorescence of the samples and standards are determined using a fluorimeter. The concentration of dsDNA in the test sample is calculated from the standard curve. The number of virus particles is calculated by the following formula:

$$\text{virus particles (VP/mL)} = \text{DNA (ng/mL)} \times \frac{\text{dilution factor}}{1000} \times \frac{\text{extinction coefficient}}{50}$$

The extinction coefficient for OAdV623 is 1.3×10^{12}.

4. To test more than two virus samples, set up a corresponding number of cell plates. However, because the health of the cells can suffer when working with large numbers of plates, the number of samples to be tested concurrently should be minimized.

5. Use a Finnpipette Biocontrol or similar electronic instrument in stepper mode with a 300 μL × 12-channel head fitted with 9 tips and set at 3 × 100 μL on "slow."

6. Use polypropylene tubes and microtiter plates when diluting as virus adsorbs to polystyrene. However, use polystyrene tubes and microtiter plates (Greiner Bio-one) for FP253 because the lipid adheres to other plastics. Dilute FP253 in TS buffer to 5×10^9 vp/mL (in a 15-mL polystyrene tube). The overage of each diluted sample should be at least 5% of the total volume required for the assay.

7. The maximum tolerated concentration of fludarabine phosphate should be determined for individual cell lines. To examine any effect of the virus alone on cells, set up test plates to which RPMI plus 20% FBS is added without fludarabine phosphate.

8. In this equation $X_0 = 0$. X_0 is only included because Origin has this equation available as an option. The equation should actually be $Y = Y_0 + Ae^{-X/t}$.

9. Host cell protein may be detected by Western transfer using a suitable antiserum generated against a CSL503 cell lysate. By spiking virus preparations with decreasing amounts of lysate, the limit of detection can be determined. Similarly, host cell DNA is detected by the amplification of a centromerically located satellite DNA sequence from the CSL503 host genome by selected primers designed to flank this area. Specialized tests for endotoxin, residual CsCl, and bio-burden must be carried out by an accredited contractor and are outside the scope of this chapter.

Acknowledgments

The authors wish to thank Dr. Minoo Moghaddam for the synthesis of CS087, Dr. Kerrie Setiawan, for suggesting the use of a linear dilution method to set up the cell standard curve, Neralie Coulston for technical assistance, Liz Kennings for providing external validation, Jane Slobedman for refining the analysis of the cell killing assay, and I. Keith Smith and Dr. Trevor Lockett for oversight and helpful discussions.

References

1. Curiel, D. and Douglas, J., eds. (2002) *Adenoviral Vectors for Gene Therapy.* Academic Press, San Diego, CA.
2. Both, G. (2002) Xenogenic adenoviruses, in *Adenoviral Vectors for Gene Therapy* (Curiel, D. and Douglas, J., eds.), Academic Press, San Diego, CA, pp. 447–479.
3. Benko, M. and Harrach, B. (1998) A proposal for a new (third) genus within the family Adenoviridae. *Arch. Virol.* **143,** 829–837.
4. Both, G. W. (2004). Ovine atadenovirus: a review of its biology, biosafety profile and application as a gene delivery vector. *Immunol. Cell Biol.* **82,** 189–195.
5. Voeks, D., Martiniello-Wilks, R., Madden, V., et al. (2002) Gene therapy for prostate cancer delivered by ovine adenovirus and mediated by purine nucleoside phosphorylase and fludarabine in mouse models. *Gene Ther.* **9,** 759–768.
6. Martiniello-Wilks, R., Dane, A., Voeks, D. J., et al. (2004) Gene-directed enzyme prodrug therapy for prostate cancer in a mouse model that imitates the development of human disease. *J. Gene Med.* **6,** 43–54.
7. Martiniello-Wilks, R., Wang, X.-Y., Voeks, D., et al. (2004) Purine nucleoside phosphorylase and fludarabine phosphate gene-directed enzyme prodrug therapy suppresses primary tumour growth and pseudo-metastases in a mouse model of prostate cancer. *J. Gene Med.* **6,** 1343–1357.
8. Wang, X.-Y., Martiniello-Wilks, R., Shaw, J., et al. (2004) Preclinical evaluation of a prostate targeted gene directed enzyme prodrug therapy delivered by ovine atadenovirus. *Gene Ther.* **11,** 1559–1567.
9. Vrati, S., Macavoy, E. S., Xu, Z. Z., Smole, C., Boyle, D. B., and Both, G. W. (1996) Construction and transfection of ovine adenovirus genomic clones to rescue modified viruses. *Virology* **220,** 200–203.
10. Löser, P., Hofmann, C., Both, G. W., Uckert, W., and Hillgenberg, M. (2003) Construction, rescue, and characterization of vectors derived from ovine atadenovirus. *J. Virol.* **77,** 11,941—11,951.
11. Löser, P., Kümin, D., Hillgenberg, M., Both, G., and Hofmann, C. (2001) Preparation of ovine adenovirus vectors. In *Methods in Molecular Medicine, Gene Therapy Protocols*, 2nd ed. (Morgan, J., ed.), Humana Press, Totowa, NJ, Vol. 69, pp. 415–426.
12. Cameron, F. H., Moghaddam, M. J., Bender, V. J., Whittaker, R. G., Mott, M., and Lockett, T. J. (1999) A transfection compound series based on a versatile tris linkage. *Biochim. Biophys. Acta* **1417,** 37–50.
13. Fasbender, A., Zabner, J., Chillon, M., et al. (1997) Complexes of adenovirus with polycationic polymers and cationic lipids increase the efficiency of gene transfer in vitro and in vivo. *J. Biol. Chem.* **272,** 6479–6489.
14. Schagen, F. H. E., Rademaker, H. J., Rabelink, M. J., et al. (2000) Ammonium sulphate precipitation of recombinant adenovirus from culture medium: an easy method to increase the total virus yield. *Gene Ther.* **7,** 1570–1574.
15. Mittereder, N., March, K. L., and Trapnell, B. C. (1996) Evaluation of the con-

centration and bioactivity of adenovirus vectors for gene therapy. *J. Virol.* **70,** 7498–7509.

16. Maizel, J. V., White, D. O., and Scharff, M. D. (1968) The polypeptides of adenovirus. I. Evidence for multiple protein components in the virion and a comparison of types 2, 7A, and 12. *Virology* **36,** 115–125.

17. Spearman, C. (1908) The method of right and wrong cases (constant stimuli) without Gauss formulae. *Br. J. Psychol.* **2,** 227–242.

18. Karber, G. (1931) Beitrag zur kollectiven Behandlung phamakalogischer Reihenversuche. *Arch. Exp. Pharmak.* **162,** 480–487.

7

Construction of Capsid-Modified Recombinant Bovine Adenovirus Type 3

Alexander N. Zakhartchouk, Qiaohua Wu, and Suresh K. Tikoo

Summary

Adenoviruses have become a popular vehicle for gene transfer into animal and human cells. However, wide prevalence of preexisting immunity to human adenovirus (HAdV) and the promiscuous nature of the virus have made the use of nonhuman adenoviruses an attractive alternative. Moreover, readministration of viral vectors is often required to maintain therapeutic levels of transgene expression, resulting in vector-specific immune responses. Although a number of features of bovine adenovirus (BAdV)-3 make it attractive for use as a vector in human vaccination, BAdV-3 transduces nonbovine cells, including human cells, poorly. However, genetic modification of capsid proteins (e.g., fiber, pIX) has helped in increasing the utility of BAdV-3 as a vector for transducing nonbovine cells. Here, we will describe the methods used to construct recombinant BAdV-3 expressing chimeric fiber or chimeric pIX proteins.

Key Words: Bovine adenovirus (BAdV)-3; pIX; fiber; tropism; gene therapy; vaccination; EYFP; human adenovirus-5.

1. Introduction

Wild-type BAdV-3 was first isolated from a healthy cow and has been shown to replicate in the respiratory tract of cattle with mild or no clinical symptoms *(1)*. Like other adenoviruses, BAdV-3 is a nonenveloped icosahedral particle of 75 nm diameter *(2)* containing a linear double-stranded DNA molecule. Although overall organization of the BAdV-3 genome appears to be similar to those of other human adenoviruses (HAdVs), the determination of the complete DNA sequence and transcriptional map of the BAdV-3 genome revealed certain distinct features of BAdV-3 *(3–8)*. The E1A region encodes three related proteins of 43, 57, and 65 kDa, and the E1B region encodes two unrelated phosphoproteins of 19($E1B_{Small}$) and 48($E1B_{Large}$) kDa *(8)*. The E3

From: *Methods in Molecular Medicine, Vol. 130:*
Adenovirus Methods and Protocols, Second Edition, vol. 1:
Adenoviruses Ad Vectors, Quantitation, and Animal Models
Edited by: W. S. M. Wold and A. E. Tollefson © Humana Press Inc., Totowa, NJ

region encodes a 69-kDa glycoprotein and a 14.5-kDa protein *(7)*. The cloning of BAdV-3 genomic DNA in a plasmid using the homologous recombination machinery of *Escherichia coli* BJ5183 *(9)* has helped to manipulate the genome quite efficiently *(10)*. In addition, transfection of in vitro manipulated BAdV-3 genomic DNA in fetal bovine retina cells (FBRC) or the VIDO R2 cell line (HAdV-5 E1-transformed bovine retina cells) *(11)* not only increased the chances of isolating the desired recombinant BAdV-3 but also eliminated the chance of isolating wild-type BAdV-3. Using this technique, we have (1) demonstrated that while E1A is essential for replication of BAdV-3 *(11)*, the E1B$_{Small}$ *(12)*, E3 *(13)*, and most of the E4 regions *(14)* are not essential for replication of the virus in Madin–Darby bovine kidney (MDBK) cells, (2) identified three sites (E1A region, E3 region, and between the right inverted terminal repeat and the E4 region) for the insertion of foreign genes *(6,13,15–17)*, and (3) identified the *cis*-acting DNA packaging domain *(18)*. In addition, we have developed, tested, and confirmed the feasibility of using BAdV-3 as a live viral vector for delivery of vaccine antigens to the respiratory mucosal surfaces of cattle *(19)*. Thus, we have established the feasibility of engineering BAdV-3 for the purpose of using it as a vaccine-delivery vehicle in cattle.

A number of features of BAdV-3 *(10,20,21)* make it attractive for use as a vector in human vaccination; however, BAdV-3 transduces nonbovine cells, including human cells, poorly *(22)*. Recently, we characterized the major (fiber) and minor (pIX) capsid proteins of BAdV-3 *(8,20,23)* and demonstrated that recombinant BAdV-3 containing a chimeric fiber (constructed by exchanging the knob region of BAdV-3 fiber with the knob region of HAdV-5 fiber) *(22)* or chimeric pIX can be isolated and used to infect and transduce nonbovine cells including human cells *(20)*.

Two cell lines have been used for generating recombinant BAdV-3s. The secondary FBRC line was established from the retina of a calf *(13)* and has been used for generating replication-competent BAdV-3s. The VIDO R2 cell line was constructed by transforming FBRC cells with human adenovirus type 5 E1 region *(11,24)* and has been used for generating both replication-competent and replication-defective (E1A-deleted) recombinant BAdV-3s. In addition, a hybrid cell line has also been isolated, which can support the replication of E1A-deleted BAdV-3 *(25)*.   .

2. Materials

1. QIAGEN Plasmid Maxi Kit (QIAGEN, Mississauga, ON, Canada, cat. no. 12162).
2. QIAprep Spin Miniprep Kit (QIAGEN, Mississauga, ON, Canada, cat. no. 27106).
3. Pfu DNA polymerase with 10X Pfu buffer (Stratagene, La Jolla, CA).

 4. Polymerase chain reaction (PCR) tubes (0.2-mL; VWR, cat. no. 20170-010).
 5. Premium flex Eppendorf 1.5-mL tubes (VWR, cat. no. CA20901-551).
 6. PCR cycler.
 7. Ultrapure agarose (Invitrogen, cat. no. 15510-027).
 8. QIAquick gel extraction kit (QIAGEN, Mississauga, ON, Canada, cat. no. 28704).
 9. Chemically competent *E. coli* strains BJ5183 *(9)* and DH5α™ (Invitrogen, cat. no. 18265-017).
10. Difco LB broth, Miller (BD, cat. no. 244620).
11. Difco LB agar, Miller (BD, cat. no. 244520).
12. Ampicillin sodium salt (Sigma, cat. no. A9518).
13. 100 × 15 mm Petri dishes (VWR, cat. no. CA25384-142).
14. LB agar plates containing 50 μg/mL ampicillin.
15. VIDO R2: HAdV-5 E1-transformed FBRC *(11)*.
16. Madin-Darby bovine kidney (MDBK) cells (ATCC CCL-22).
17. FBRC *(13)*.
18. Costar 6-well cell culture plates (Fisher, cat. no. CS003506); Costar 24-well cell culture plates (Fisher, cat. no. CS003524); Costar 12-well cell culture plates (Fisher, cat. no. CS003513); Costar 96-well cell culture plates (Fisher, cat. no. CS003596); Corning 150-cm^2 cell culture flasks (Fisher, cat. no. 1012632); Corning 75-cm^2 cell culture flasks (Fisher, cat. no. 1012631); Corning 25-cm^2 flasks (Fisher, cat. no. 1012630).
19. Disposable cell scrapers (Fisher, cat. no. 08-773-2).
20. Corning 50-mL centrifuge tubes (Fisher, cat. no. 05-538-55A).
21. 175-mL Centrifuge bottles (VWR, cat. no. 21008-942).
22. 13 × 51 mm Ultra-Clear centrifuge tubes (Beckman, cat. no. 344057).
23. 14 × 89 mm Ultra-Clear centrifuge tubes (Beckman, cat. no. 344059).
24. Slide-A-Lyser Dialysis Cassette (Biolynx Inc., cat. no. 66425).
25. Eagle's minimum essential medium (MEM; Invitrogen, cat. no. 11095-080, Burlington, ON, Canada).
26. Fetal bovine serum (Sera Care Life Sciences Inc., cat. no. CC-5010).
27. (1X) Trypsin-ethylene diamine tetraacetic acid (EDTA) (Invitrogen, cat. no. 25300-062).
28. Lipofectin reagent (Invitrogen, cat. no. 18292052).
29. OPTI-MEM I reduced-serum medium (Gibco/Invitrogen, cat. no. 31985).
30. 2X HEBS (0.5 L): dissolve in dionized water 5 g *N*-2-hydroxyethylpiperazine-*N*'-2-ethanesulfonic acid (HEPES), 8 g NaCl, 0.37 g KCl, 0.125 g Na$_2$HPO$_4$ dihydride, 1 g glucose (all reagents are from Sigma, tissue-culture grade). It is critical that the pH of the solution be between 6.95 and 6.98. Filter-sterilize through a 0.45-μm nitrocellulose filter (Nalgene) and store in aliquots at –20°C.
31. 2.5 *M* CaCl$_2$: add 183.7 g of CaCl$_2$ dihydride (Sigma, tissue-culture grade) to 0.5 L of water. Filter-sterilize through a 0.45-μm nitrocellulose filter (Nalgene) and store in aliquots at –20°C.
32. TE buffer: 10 m*M* Tris-HCl, pH 7.5, 1 m*M* EDTA.

33. TE-saturated phenol (Sigma, cat. no. 77607).
34. Chloroform (Sigma, cat. no. I9516).
35. Phosphate-buffered saline (PBS), pH 7.4: 58 mM Na$_2$HPO$_4$, 17 mM NaH$_2$PO$_4$, 68 mM NaCl.
36. 5 M NaCl.
37. 1 M MgCl$_2$.
38. 10% SDS.
39. Triton X-100 (Sigma, X100RS).
40. Salted ethanol (2% potassium acetate in pure ethanol).
41. 70% Ethanol.
42. Proteinase K treatment solution: 10 mM Tris-HCl, pH 7.8, 5 mM EDTA, 0.5% sodium dodecyl sulfate (SDS), 100 µg/mL Proteinase K.
43. RNase (10 mg/mL).
44. Complete MEM: Add to one 0.5 L bottle of MEM with Earl's salts and L-glutamine (Invitrogen, cat. no. 11095-080), 5 mL MEM nonessential amino acids solution (10 mM; Invitrogen, cat. no. 11140-050), 5 mL HEPES buffer solution (1 M; Invitrogen, cat. no. 15630-080) and 0.5 mL gentamicin reagent solution (Invitrogen, cat. no. 15750-060).
45. 5% Sodium deoxycholate.
46. Deoxyribonuclease I (10 MU) (MP Biomedicals, cat. no. 190062). Make 10 mg/mL solution of deoxyribonuclease I (10 MU) in sterile water. Aliquot and store at –20°C.
47. CsCl solutions for banding: verify density by weighing 1 mL of each solution. Sterilize by filtration and store at room temperature.
    a. 1.35 g/mL: prepare by adding together 70.4 g of CsCl and 129.6 g of 10 mM Tris, pH 8.0.
    b. 1.25 g/mL: prepare by adding together 54.0 g of CsCl and 146.0 g of 10 mM Tris, pH 8.0.
48. Slide-A-Lyzer 10K dialysis cassette (Pierce).
49. Sigma Type VII low gelling temperature agarose (Sigma, cat. no. A-9045). To prepare 1.4% low melting agarose, add 1.4 g agarose to 100 mL of water. Autoclave, aliquot in 50 mL sterile tubes with caps, store at 4°C.
50. Extraction solution: 400 mM NaCl, 50 mL/L Triton X-100, and 50 mM Tris-HCl, pH 7.5.
51. Rabbit anti-DNA binding protein (DBP) polyclonal antibody.
52. Biotinylated anti-rabbit IgG from a VECTASTAIN Rabbit IgG ABC kit (Vector Laboratories).
53. Streptavidin–horseradish peroxidase (HRP) conjugate.
54. DAB peroxidase substrate kit (Vector Laboratories).

3. Methods

3.1. Construction of Transfer Plasmids

3.1.1. Construction of Transfer Plasmid Containing Chimeric Fiber by Overlap Extension PCR

Although we are describing here the construction of chimeric fiber in which the knob region of BAdV-3 fiber has been replaced with the knob region of HAdV-5 fiber *(11)* (**Fig. 1A**), the method used for creating chimeric fiber should be applicable to replacement of the BAdV-3 knob with any other adenovirus fiber knob.

1. Purify plasmid DNA containing the gene encoding BAdV-3 *(5)* or HAdV-5 fiber *(24)* using Plasmid Maxi Kit (QIAGEN) according to the manufacturer's instructions and determine the concentration of purified DNA.
2. PCR-amplify the DNA fragment encoding the BAdV-3 fiber shaft domain and the DNA fragment encoding the HAdV-5 fiber knob domain and the polyadenylation (poly[A]) using purified plasmid DNA as templates and the appropriate primers (*see* **Note 1**). Set up the PCR mixture in a thin-well 0.2-mL tube as follows: 2 µL of plasmid DNA (10 ng/µL), 3 µL of dNTPs (10 m*M* each), 2.5 µL of each primer (10 p*M*/µL), 5 µL of 10X Pfu buffer, 1 µL of Pfu DNA polymerase (2.5 U/µL; Stratagene), and 34 µL of H_2O. Run PCR reaction with 5 min of initial denaturation at 95°C, 25 cycles of 1-min denaturation at 94°C, 1 min annealing at 55°C, 1 min 30 s extension at 72°C, and finally 7 min extension at 72°C.
3. Load PCR reactions (total of 50 µL each) onto 0.8% agarose gel and check the PCR products by electrophoresis.
4. Purify the PCR products from the gel using QIAquick Gel Extraction Kit (QIAGEN) based on manufacturer's instructions.
5. Mix PCR products together and perform second PCR reaction with primers A and D (**Fig. 1A**).
6. Purify the PCR product as described in **steps 3** and **4** (*see* **Note 2**).
7. Digest the PCR product with restriction enzymes *Eco*RI and *Mun*I and clone into the *Eco*RI site of the DNA plasmid containing the BAV-3 fiber gene to generate the plasmid pGSAd5FK.
8. Digest the plasmid pGSAd5FK with *Rsr*II, isolate a 2.5-kb DNA fragment, and clone into *Rsr*II-digested plasmid pBAV-301.gfp *(11,22)* to generate plasmid pBAV-301.G5FK (**Fig. 1B**).

3.1.2. Construction of Transfer Plasmid Containing Chimeric pIX Gene

We have used the *Hpa*I site (nt 3560 of BAdV-3 genome) in the 3'-end of the pIX gene to incorporate heterologous sequences into the BAdV-3 pIX C-terminus. This has resulted in the removal of the last four amino acids of the wild-type BAdV-3 pIX. Although we have depicted here the scheme for con-

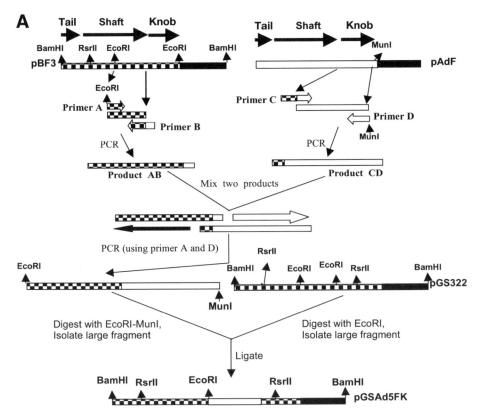

Fig. 1. Schematic diagram depicting the construction of (A) transfer vector for chimeric BAdV-3 fiber and (B) full-length BAdV-3 genomic DNA clone (pFBAV600) containing chimeric fiber. Plasmid DNA (■■), BAdV-3 genomic DNA (▨▨), HAdV-5 fiber DNA (☐), GFP DNA (IIIII), human cytomegalovirus immediate early promoter (◩), bovine growth hormone polyA (◓). (Adapted from **ref. 22**.)

struction of chimeric pIX in which the C-terminus of BAdV-3 pIX is fused to the enhanced green fluorescent protein gene (EYFP) (**Fig. 2**), the method and vectors used for creating chimeric pIX should be applicable to fusion of BAdV-3 pIX to any other ligand.

3.2. Construction of Full-Length Plasmids by Homologous Recombination in E. coli

1. Isolate a *KpnI-SwaI* DNA fragment from plasmid pBAV-301.G5FK (**Fig. 1B**) or *PacI-NotI* fragment of plasmid pBAVNotYFP (**Fig. 2**) using QIAquick Gel Extraction kit.
2. Mix the *KpnI-SwaI* DNA fragment with *SrfI*-digested plasmid pFBAV-302 (E3-deleted BAdV-3 DNA; (*13,22*)) or the *PacI-NotI* fragment with *BsaBI-PmeI*

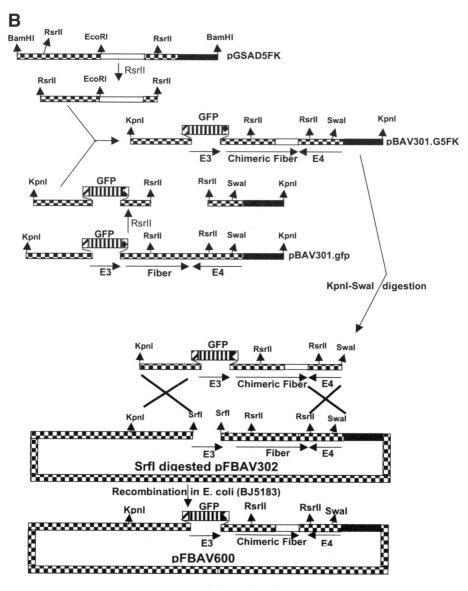

Fig. 1. (*continued*)

digested plasmid pFBAV3 *(20)* (*see* **Note 3**). Add the individual mixtures into 100 µL of chemically competent *E. coli* BJ5183 and incubate on ice for 30 min. Heat-shock the cells for 1 min at 42°C without shaking. Immediately transfer the tube to ice. Add 100 µL of room temperature LB broth. Plate the whole sample onto a prewarmed LB–ampicillin agar plate and incubate overnight at 37°C (*see* **Note 4**).

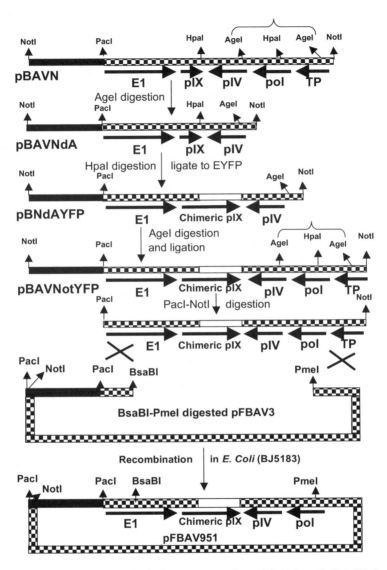

Fig. 2. Schematic diagram depicting construction of full-length BAdV-3 genomic DNA clone (pFBAV951) containing chimeric pIX. Plasmid DNA (■■), BAdV-3 genomic DNA (▨▨), EYFP (▭). (Adapted from **ref. 20**.)

3. Pick 20 of the individual colonies and grow them overnight in 1.5 mL of LB broth containing 50 µg/mL ampicillin.
4. Extract plasmid DNA by the alkaline lysis method *(27)*, dissolve the DNA in 5 µL of H₂O, and chemically transform the whole plasmid DNA preparation into *E. coli* DH5α as described (*see* **Note 5**).

5. Pick and grow isolated colonies (*see* **Note 6**) in 1.5 mL of LB broth containing 50 µg/mL ampicillin, and extract DNA by the alkaline lysis method *(27)*. Screen the full-length plasmid DNA with appropriate restriction enzymes to confirm the presence of the chimeric fiber or chimeric pIX.

6. Purify full-length plasmid DNA that contains chimeric fiber or chimeric pIX gene using Plasmid Maxi Kit (QIAGEN) according to manufacturer's instruction and determine the concentration of the purified DNA.

3.3. Construction and Isolation of Capsid Modified BAdV-3s

1. Grow VIDO R2 cells overnight to 80–85% confluency in 6-well plates in MEM supplemented with 10% FBS. Alternatively, grow FBRC cells overnight to 60% confluency in 6-well plates in MEM supplemented with 10% serum. Before transfection, wash the cells once with 2 mL of OPTI-MEM I reduced-serum medium (*see* **Note 7**).

2. Digest 40 µg of individual plasmid DNA (**Subheading 3.2., step 6**) with *Pac*I. Extract digested plasmid DNA once with phenol/chloroform. Precipitate DNA with pure ethanol and resuspend DNA in 100 µL of sterile distilled H_2O. Check the concentration of the DNA.

3. For transfection of VIDO R2 cells grown in one well of a 6-well plate, mix 10 µL of Lipofectin with 50 µL of OPTI-MEM I reduced-serum medium. Incubate the sample for 30 min at room temperature. Dissolve 5–10 µg (*see* **Note 8**) of the *Pac*I-digested pFBAV-600 recombinant plasmid DNA in a total of 50 µL of OPTI-MEM I reduced-serum medium. Add 60 µL of Lipofectin–OPTI–MEM I reduced-serum medium into the DNA solution and incubate the sample for 30 min at room temperature. Add 0.9 mL of OPTI-MEM I reduced-serum medium into the above Lipofectin–DNA mixture and slowly add to the VIDO R2 cell monolayer.

4. For transfection of FBRC cells, set up 6-well plate of FBRC cells the day before a transfection in 3 mL of MEM + 10% FBS per well. The cells need to be 70–80% confluent at the time of transfection. In a sterile 1.5-mL microcentrifuge tube, mix 6 µg of the DNA (*see* **Note 8**) and sterile deionized water in a total volume of 135 µL. Add 150 µL of 2X HEBS (*see* **Heading 2., item 30, and Note 9**) and mix well. Add 15 µL of 2.5 *M* $CaCl_2$ and vortex thoroughly (*see* **Heading 2., item 31**, and **Note 10**). Incubate at room temperature for 30 min to allow a precipitate to form. Add the calcium phosphate–DNA solution drop wise to the medium in the well while swirling the plate.

5. Incubate the cells for 6 h at 37°C in a CO_2 incubator. Then replace the medium with MEM supplemented with 3% FBS. Change the medium when necessary and check the transfected cells every other day for the presence of viral plaques (*see* **Note 11**).

6. When 50% of the transfected cells show cytopathic effects (CPE), scrape the cells into the medium and pellet them for 10 min at 600*g*. Resuspend the cells in 1 mL of fresh MEM. Perform three freeze–thaw cycles (–70°C and 37°C). The virus stock should be labeled as passage 1 and stored at –70°C.

3.4. Analysis of Potential Recombinant Viruses

3.4.1. Analysis of Virion DNA

1. Set up MDBK cells in one T75 flask and infect cells on the next day with 100 µL of recombinant virus stock (**Subheading 3.3., step 6**).
2. Two days later, when more than 50% of cells show CPE, extract the viral DNA from infected cells by the modified Hirt's method *(26)*. Scrape the cells and pellet them by low-speed centrifugation. Aspirate culture medium and resuspend the cell pellet in 100 µL of PBS.
3. In a 1.5-mL Eppendorf tube, mix 100 µL of the cell suspension and 200 µL of the extraction solution: 400 mM NaCl, 50 mL/L Triton X-100, and 50 mM Tris-HCl, pH 7.5.
4. Incubate at 4°C for 30 min with shaking.
5. Spin 5000 rpm (2320g) in an Eppendorf centrifuge, 15 min at 4°C.
6. Transfer the supernatant to a fresh tube.
7. Add 30 µL 10% SDS and 20 µL 10 mg/mL Proteinase K.
8. Incubate at 37°C for 1 h.
9. Extract twice with phenol and once with chloroform. Transfer the solution from the aqueous phase to a fresh tube.
10. Precipitate the DNA with two volumes of salted ethanol for 1 h at –20°C.
11. Centrifuge the sample at 27,000g for 30 min at 4°C. Dissolve the DNA pellet in 30 µL of water containing 125 µg/mL of RNase.
12. The extracted recombinant viral DNA can be analyzed by restriction enzyme digestion, Southern blot *(27)*, and PCR (using specific primers) to verify the overall size of the viral genome and to confirm the presence of the gene encoding the chimeric protein.

3.4.2. Analysis of Chimeric Proteins

The identity of the recombinant BAdV-3s can also be confirmed by detecting the expression of chimeric proteins in virus-infected cells by Western blot analysis *(28)* or by immunoprecipitation assay *(29)* using protein-specific antibodies (*see* **Note 12**).

3.5. Plaque Purification

1. Set up MDBK cells in 6-well plates so that they form 60–70% confluent monolayers the next day.
2. Prepare serial 10-fold dilutions as described in **Subheading 3.8.1.**
3. Aspirate medium from the wells and add 0.5 mL of the diluted virus (2 wells per each dilution starting from 10^4). Incubate at 37°C for 1 h.
4. Aspirate virus solution and overlay the cells with agarose as described in **Subheading 3.8.2.** Add 3 mL of agarose/medium mix per well. Incubate at 37°C in a CO_2 incubator for 10–14 d.
5. Check for the presence of plaques in the monolayer under a microscope. Transfer single plaques from the highest dilution containing plaque(s) using a pipetman

and sterile plugged tip(s) to a sterile Eppendorf tube with 0.2 mL of culture medium.

6. Freeze the tube(s) at –70°C.

3.6. Making a Virus Stock From a Plaque

1. Infect a 60–70% confluent MDBK cell monolayer in a 24-well plate with 0.1 mL thawed plaque material and incubate at 37°C until 100% of cells show CPE.

2. When the cells have significant CPE, harvest them by scraping the cells into the medium with a cell scraper. Freeze–thaw the cells three times, then centrifuge 5 min at 2000g to remove the cellular debris. Collect the supernatant and store virus stock at –70°C.

3. Using the newly obtained virus stock and, following the directions given above (**Subheading 3.6., step 1**), infect one well of a 6-well plate (70–80% confluent monolayer of MDBK cells) with 0.1 mL crude virus per well. Harvest as above and infect T25 flask of 60–70% confluent MDBK cells monolayer (0.2 mL per flask). Harvest as above and infect T75 flask with a 70–80% confluent MDBK cells monolayer (1 mL per flask). Harvest as above and infect a T150 flask with a 70–80% confluent MDBK cells monolayer (2 mL per flask).

3.7. CsCl Gradient Purification

1. Set up 20 T150 flasks of MDBK cells per recombinant virus. The cells need to be 70–80% confluent at the time of infection.

2. Aspirate the medium from each flask and add 1 mL of virus plus 4 mL of serum-free MEM per flask (*see* **Subheading 2., item 44**). Rock the flasks to ensure that all cells are covered by the medium. Incubate at 37°C for 1 h, then add 20 mL of MEM containing 5% FBS. Monitor the cells for CPE and harvest when most of the cells are rounded. Harvest the cells by scraping them into the medium, and collect them by centrifugation for 15 min at 750g in 175-mL conical tubes. Resuspend the cell pellet in 15 mL of PBS. Store the sample at –70°C.

3. Thaw sample and add 1.5 mL 5% sodium deoxycholate. Mix well and incubate at room temperature for 30 min. Add 0.3 mL 1 M MgCl$_2$ and 75 µL DNase I solution (*see* **Subheading 2., item 46**). Mix well and incubate at 37°C for 30 min, mixing every 10 min (*see* **Note 13**). Centrifuge at 2000g for 30 min at 4°C.

4. Meanwhile, prepare a CsCl step gradient in SW41 ultraclear tubes. Add 4 mL of 1.25 g/mL solution and then gently underlay 1.5 mL of 1.35 g/mL solution (*see* **Subheading 2., item 47**).

5. Apply 6 mL of supernatant from **step 3** to the top of each gradient. If necessary, top off tubes with PBS.

6. Spin at 35,000 rpm (209,490g) in SW41 rotor, 4°C for 2 h.

7. Collect the lower band and pool (*see* **Note 14**).

8. Transfer the pooled virus to a SW50.1 Ultraclear tube. Use the 1.35 g/mL CsCl solution to top off the tube and mix well.

9. Spin at 35,000 rpm (146,821g) in the SW50.1 rotor, 4°C for 16–20 h.

10. Collect the virus band in a small volume (usually 0.5–1.0 mL), transfer to a Slide-A-Lyzer 10K dialysis cassette (Pierce), and dialyze at 4°C against three changes of 1.5 L 10 mM Tris-HCl, pH 8.0, overnight.
11. After dialysis, add sterile glycerol to a final concentration of 10%. Store the purified virus at –70°C in small aliquots.

3.8. Quantitation of Virus by Titration

3.8.1. TCID$_{50}$ Assay

1. Trypsinize a monolayer of MDBK cells in a T150 flask.
2. Add 100 µL of cell suspension (5×10^5 cells/mL in MEM plus 10% FBS) in each well of a 96-well plate.
3. Place the plate into an incubator. Meanwhile, thaw an aliquot of virus stock.
4. Aliquot 1.8 mL of serum-free MEM into each of 10 small glass test tubes with caps and number and order these tubes from 1 to 10.
5. Add 0.2 mL of virus to tube 1. Vortex and remove 0.2 mL from tube 1 and deposit it into tube 2. Vortex and repeat for the remaining tubes, creating a 10-fold dilution series of the original virus stock.
6. Take the 96-well plate with MDBK cells from the incubator, and label vertical rows with dilution numbers. Leave one mock row.
7. Add 0.1 mL of serum-free MEM to the 8 wells of the mock row.
8. Beginning with the highest dilution, add 0.1 mL of the diluted virus to each of the 8 wells of the row.
9. Incubate at 37°C in a CO_2 incubator for 10–14 d.
10. Monitor the wells for the presence of CPE under a microscope.
11. The calculation formula is $V = X_0 - 1/2 + (1/8 \times \Sigma X_i)$, where X_0 is the highest dilution number where each of the 8 tested wells shows CPE; ΣX_i is the sum of all the wells that give CPE counting from the X_0 row, so 8 plus the rest of the wells gives CPE after that. Finally, TCID$_{50}$/mL $= 10^V \times 10$, where 10 is a dilution factor.

3.8.2. Determination of Infectious Units by DNA Binding Protein Assay (30)

1. Set up 12-well plates of MDBK cells to be 60–70% confluent at the time of infection. Prepare a 10-fold dilution series of the virus stock as described in **Subheading 3.8.1.**, **steps 4** and **5**.
2. Add 0.5 mL of the solution to each well of a 3-well row, and add MEM to one mock-infected well. Incubate at 37°C for 1 h.
3. Meanwhile, microwave-sterilize 1.4% agarose (*see* **Subheading 2.**, **item 49**) until it is boiling, and place it in a 40°C water bath. Allow agarose to equilibrate at 40°C for at least 30 min. Warm 2X MEM containing 5% FBS at 37°C.
4. After the 1 h incubation, aspirate virus from the wells. Mix equal volumes of warm 2X MEM and agarose, and, working quickly, overlay each well with 2 mL of medium/agarose by gently applying the solution to the cells with a pipet, touching the pipet tip to the inner wall of the well. Start overlaying with the mock well, and then continue with the highest virus dilution. Try not to introduce bubbles onto the well.

5. Leave the plate for 10–15 min at room temperature to let the agarose solidify. Place the plate into an incubator at 37°C.
6. Three days later, place the plate at 4°C for 1 h.
7. Remove the agarose from the wells using a spatula and wash cells with PBS.
8. Fix the cells by adding 0.5 mL of acetone/methanol mix. Incubate at –20°C for 15 min.
9. Wash the cells with PBS and add rabbit anti-DBP polyclonal antibodies (dilution is 1:100, 0.5 mL per well). Incubate for 2 h at room temperature.
10. Wash the cells with PBS and add biotinylated anti-rabbit IgG from a VECTASTAIN Rabbit IgG ABC kit (Vector Laboratories): 1 drop of the stock to 10 mL of PBS. Incubate for 1 h at room temperature.
11. Wash the cells with PBS and add streptavidin-HRP conjugate. For 10 mL PBS add two drops of reagent A and two drops of reagent B from the VECTASTAIN kit. Incubate 30 min at room temperature.
12. Wash the cells with PBS and stain them using a DAB Peroxidase Substrate kit (Vector Laboratories).
13. Wash the cells with PBS and store the plates at 4°C.
14. Count the positive cells (dark brown) under a microscope. The number of positive cells per well multiplied by a dilution number and dilution factor (2) gives the titer of the virus in DBP units per mL.

4. Notes

1. Design the primers according to the BAdV-3 and HAdV-5 sequences from the National Center for Biotechnology Information (BAdV-3 GenBank accession no. AF030154; HAdV-5 GenBank accession no. M73260).
2. The size of the chimeric fiber gene can be estimated. The chimeric fiber gene sequence should be verified by DNA sequencing before proceeding to the next step.
3. We usually check the quality of the DNA by agarose gel electrophoresis. Usually, we mix 100 ng of each of DNA fragment and the plasmid DNA. Although we have treated plasmid DNA with alkaline phosphatase occasionally, it does not appear to be essential for getting positive DNA clones after recombination in BJ5183.
4. We usually get between 10 and 50 colonies after chemical transformation. To increase the chances of getting more colonies, electroporation can also be used for transformation of *E. coli* BJ5183 cells.
5. The yield and quality of plasmid DNA in *E. coli* BJ5183 cells are poor (*9*). Therefore, amplification of the plasmid DNA in the DH5α strain is required to obtain sufficient plasmid DNA for restriction enzyme analysis.
6. We usually pick about 20 colonies for further analysis. We have found that some colonies may not grow in LB–ampicillin broth and plasmid DNA extracted from some colonies will not give the expected DNA fragments after restriction enzyme analysis. Therefore, 20 colonies may be an optimal number to start.
7. VIDO R2 cells are grown in MEM supplemented with 10% FBS. The cells should be plated in a 6-well plate so that the cell reaches about 50–60% confluency on the day of transfection.

8. To increase the chance of isolating recombinant BAdV-3, we have usually used three concentrations (5, 7.5, and 10 µg) of DNA/well of a 6-well plate per plasmid.

9. It is critical that the pH of the solution be between 6.95 and 6.98. The solution can be frozen and thawed repeatedly, but it is important to warm it to room temperature prior to transfection. Alternatively, 2X HEPES-buffered saline from Promega ProFection Mammalian Transfection System kit (cat. no. E1200) can be used.

10. The final concentration of $CaCl_2$ in the transfection mix is 0.25 *M*.

11. For isolation of recombinant BAdV-3s, it takes 2–4 wk for formation of viral plaques. You can keep transfected FBRC cells for over a month without any splitting. If no viral plaques are seen in VIDO R2 after 1 wk of transfection, cells should be trypsinized and transferred to a T75 flask. We usually observe transfected cells for 4–5 wk for the development of cytopathic effects.

12. Rabbit anti-pIX M1 sera recognize BAdV-3 pIX *(20)*; GFP monoclonal antibodies (BD, cat. no. 632375) recognize EYFP; rabbit anti-fiber knob 99-44 sera recognize BAdV-3 fiber shaft *(22)*; monoclonal antibody 1D6.14 recognizes HAdV-5 knob *(31)*. We usually use 1:200 and 1:1000 dilutions of polyclonal or monoclonal antibodies, respectively.

13. After adding deoxycholate, the solution becomes viscous. After DNase treatment, the solution viscosity is reduced to only slightly greater than that of water, and it has a white color.

14. We collect a visible virus band by piercing the tube wall a few millimeters beneath the band using a needle and a syringe.

Acknowledgments

We are indebted to all postdoctoral fellows, graduate students, and technicians who have worked in the laboratory over the years and helped to develop many of the methods described here. We thank Tess Laidlaw for editorial assistance. The work reported here was supported by grants from NSERC Canada, Canadian Institute of Health Research, and Saskatchewan Health Research Foundation to S.K.T.

References

1. Mattson, D. E., Norman, B. B., and Dunbar, J. R. (1988) Bovine adenovirus type-3 infection in feedlot calves. *Am. J. Vet. Res.* **49,** 67–69.

2. Ziiyama, Y., Igarashi, K., Tsukamoto, K., Kurokawa, T., and Sugino, Y. (1975) Biochemical studies on bovine adenovirus type 3. I. Purification and properties. *J. Virol.* **16,** 621–633.

3. Lee, J. B., Baxi, M. K., Idamakanti, N., et al. (1998) Genetic organization and DNA sequence of early region 4 of bovine adenovirus type 3. *Virus Genes* **17,** 99–100.

4. Baxi, M., Reddy, P. S., Zakhartchouk, et al. (1998) Characterization of bovine adenovirus type 3 early region 2B. *Virus Genes* **16,** 313–316.

5. Reddy, P. S., Idamakanti, N., Zakhartchouk, A. N., et al. (1998) Nucleotide sequence, genome organization, and transcription map of bovine adenovirus type 3. *J. Virol.* **72,** 1394–1402.

6. Baxi, M. K., Babiuk, L. A., Mehtali, M., and Tikoo, S. K. (1999) Transcription map and expression of bovine herpesvirus-1 glycoprotein D in early region 4 of bovine adenovirus-3. *Virology* **261,** 143–152.

7. Idamakanti, N., Reddy, P. S., Babiuk, L. A., and Tikoo, S. K. (1999) Transcription mapping and characterization of 284R and 121R proteins produced from early region 3 of bovine adenovirus type 3. *Virology* **256,** 351–359.

8. Reddy, P. S., Chen, Y., Idamakanti, N., Pyne, C., Babiuk, L. A., and Tikoo, S. K. (1999) Characterization of early region 1 and pIX of bovine adenovirus-3. *Virology* **253,** 299–308.

9. Chartier, C., Degryse, E., Ganter, M., Dieterle, A., and Mehtali, M. (1996) Efficient generation of recombinant adenovirus vectors by homologous recombination in Escherichia coli. *J. Virol.* **70,** 4805–4810.

10. Rasmussen, U. B., Benchaibi, M., Meyer, V., Schlesinger, Y., and Schughart, K. (1999) Novel human gene therapy vectors: Evaluation of wild-type and recombinant animal adenoviruses in human derived cells. *Hum. Gene Ther.* **10,** 2587–2599.

11. Reddy, P. S., Idamakanti, N., Chen, Y., et al. (1999) Replication-defective bovine adenovirus type 3 as an expression vector. *J. Virol.* **73,** 9137–9144.

12. Zhou, Y., Reddy, P. S., Babiuk, L. A., and Tikoo, S. K. (2001) Bovine adenovirus type 3 E1B (small) protein is essential for growth in bovine fibroblast cells. *Virology* **288,** 264–274.

13. Zakhartchouk, A. N., Reddy, P. S., Baxi, M., et al. (1998) Construction and characterization of E3-deleted bovine adenovirus type 3 expressing full-length and truncated form of bovine herpesvirus type 1 glycoprotein gD. *Virology* **250,** 220–229.

14. Baxi, M. K., Robertson, J., Babiuk, L. A., and Tikoo, S. K. (2001) Mutational analysis of early region 4 of bovine adenovirus type 3. *Virology* **290,** 153–163.

15. Baxi, M. K., Deregt, D., Robertson, J., Babiuk, L. A., Schlapp, T., and Tikoo, S. K. (2000) Recombinant bovine adenovirus type 3 expressing bovine viral diarrhea E2 induces an immune response in cotton rats. *Virology* **278,** 234–243.

16. Reddy, P. S., Idamakanti, N., Zakhartchouk, L. N., Babiuk, L. A., Mehtali, M., and Tikoo, S. A. (2000) Optimization of bovine coronavirus hemagglutinin-estrase glycoprotein expression in E3 deleted bovine adenovirus-3. *Virus Res.* **70,** 65–73.

17. Tikoo, S. K. (2001) Animal adenovirus as a live vaccine vector. *Recent Res. Dev. Virol.* **3,** 495–503.

18. Xing, L., Zhang, L., Van Kessel, J., and Tikoo, S. K. (2003) Identification of cis-acting sequences required for selective packaging of bovine adenovirus type 3 DNA. *J. Gen. Virol.* **84,** 2947–2956.

19. Zakhartchouk, A. N., Pyne, C., Mutwiri, G. K., et al. (1999) Mucosal immunization of calves with recombinant bovine adenovirus-3: induction of protective immunity to bovine herpesvirus-1. *J. Gen. Virol.* **80,** 1263–1269.

20. Zakhartchouk, A. N., Connors, W., van Kessel, A., and Tikoo, S. K. (2004) Bovine adenovirus type 3 containing heterologous protein in the C-terminus of minor capsid protein pIX. *Virology* **320,** 291–300.
21. Moffatt, S., Hays, J., HogenEsch, H., and Mittal, S. K. (2000) Circumvention of vector-specific neutralizing antibody response by alternating use of human and non-human adenoviruses: implications in gene therapy. *Virology* **272,** 159–167.
22. Wu, Q. and Tikoo, S. K. (2004) Altered tropism of recombinant bovine adenovirus type 3. *Virus Res.* **99,** 9–15.
23. Wu, Q., Chen, Y., Kulshreshtha, V., and Tikoo, S. K. (2004) Characterization and nuclear localization of the fiber protein encoded by the late region 7 of bovine adenovirus type 3. *Arch. Virol.* **149,** 1783–1799.
24. Chroboczek, J., Bieber, F., and Jacrot, B. (1992) The sequence of the genome of adenovirus type 5 and its comparison with the genome of adenovirus 2. *Virology* **186,** 280–285.
25. van Olphen, A. L. and Mittal, S. K. (2002) Development and characterization of bovine x human hybrid cell lines that efficiently support the replication of both wild-type bovine and human adenoviruses and those with E1 deleted. *J. Virol.* **76,** 5882–5892.
26. Hirt, B. (1967) Selective extraction of polyoma DNA from infected mouse cell cultures. *J. Mol. Biol.* **26,** 365–369.
27. Sambrook, J., Fritsch, E. F., and Maniatis, T. (1989) *Molecular Cloning: A Laboratory Manual*, 2nd ed., Cold Spring Harbor Laboratory, Cold Spring Harbor, NY.
28. Gallagher, S. R., Winston, S. E., Fuller, S. A., and Hurrell, G. R. J. (2005) Immunoblotting and immunodetection, in *Current Protocols in Molecular Biology* (Ausubel, F. M., Brent, R., Kingston, R. E., et al., eds.), Wiley & Sons, New York, pp. 10–18.
29. Bonifacino, J. S., Dell'Angelica, E. C., and Springer, T. A. (2005) Immunoprecipitation, in *Current Protocols in Molecular Biology* (Ausubel, F. M., Brent, R., Kingston, R. E., et al., eds.), Wiley & Sons, New York, pp. 10–16.
30. Zhou, Y., Pyne, C., and Tikoo, S. K. (2001) Determination of bovine adenovirus-3 titer based on immunohistochemical detection of DNA binding protein in infected cells. *J. Virol. Meth.* **94,** 148–154.
31. Douglas, J. T., Rogers, B. E., Rosenfeld, M. E., Michael, S. I., Feng, M., and Curiel, D. T. (1996) Targeted gene delivery by tropism-modified adenovirus vectors. *Nat. Biotechnol.* **14,** 1574–1578.

8

Adenovirus Capsid Chimeras

Fiber Terminal Exon Insertions/Gene Replacements in the Major Late Transcription Unit

Jason Gall, John Schoggins, and Erik Falck-Pedersen

Summary

The adenovirus major late transcription unit (MLTU) encodes the main structural capsid proteins. Expression from the MLTU is accomplished through alternative mRNA processing and use of a terminal exon coding strategy. The capsid proteins hexon, penton, and fiber contribute to efficient infection by adenovirus, and each contributes in some manner to the antiviral immune response against adenovirus infection. The ability to manipulate these genes affords one the opportunity to "detarget" adenovirus, to retarget adenovirus, and to alter immune recognition. In this chapter, we are presenting a terminal exon-replacement strategy that can be used to genetically manipulate capsid proteins expressed from the MLTU. An emphasis will be placed on manipulations of fiber as an intact terminal exon.

Key Words: Capsid structural protein; terminal exon; recombinant adenovirus vector; homologous recombination; major late transcription unit; cytopathic effect.

1. Introduction

Adenovirus (Ad) gene transfer vectors offer several experimental benefits that are useful for a variety of in vitro and in vivo applications. The virtues of Ad vectors include the efficiency of gene transduction both in vitro and in vivo, the relatively large foreign DNA-carrying capacity of the virus, the ease of vector construction, and finally the ease of vector production. The serotype 5 Ad is the workhorse of Ad vectors, largely on the basis of convenience and the preexisting use of Ad5 in the majority of laboratory studies that characterized the biology of Ad infection. Ad5 is a subgroup C Ad and represents one of more than 50 known human serotypes from six subgroups (A–F).

From: *Methods in Molecular Medicine, Vol. 130:*
Adenovirus Methods and Protocols, Second Edition, vol. 1:
Adenoviruses Ad Vectors, Quantitation, and Animal Models
Edited by: W. S. M. Wold and A. E. Tollefson © Humana Press Inc., Totowa, NJ

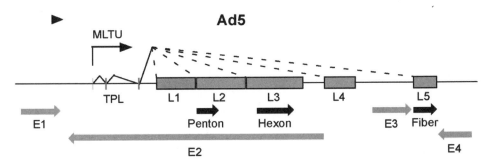

Fig. 1. Diagrammatic representation of the adenovirus type 5 genome from 0 to 100 map units (365 bp/mu). The major late transcriptional unit (MLTU) initiates transcription at 16.5 map units. Three small exons are spliced together to form the 5' tripartite leader (TPL), which, when spliced to a terminal exon from L1 to L5, generates a coding mRNA transcript.

Ad5-mediated gene transduction occurs through a well-characterized multistep process that includes high-affinity binding of capsid protein fiber to the cellular receptor Coxsackie adenovirus receptor (CAR) *(1)*. High-affinity binding is followed by a secondary interaction between $\alpha v \beta$ integrins present on the cell membrane and the Ad capsid protein penton base. Binding of integrins by Ad stimulates endocytosis and virus entry *(2)*. This entry pathway has been established through studies using cell lines and contributes to the high efficiency of Ad-mediated gene transduction. When Ad5 vectors are used to deliver a target gene in vivo, additional host factors interact with viral capsid proteins, blurring the specificity of virus infection *(3)*. Genetic manipulation of Ad capsid proteins affords a simple strategy to study capsid protein interactions, the influence of capsid proteins on virus-specific targeting, and the influence of capsid proteins on activation of the antiviral immune response.

1.1. Structural Capsid Proteins Expressed From the Major Late Transcription Unit

Adenovirus infections are temporally defined based on an ordered progression of gene transcription. In the majority of vectors, the immediate early gene E1A and two early regions, E1B and E3, have been deleted. Following early gene expression is the onset of DNA replication, which coincides with late gene expression. All viral capsid proteins are expressed during the late stage of viral infection and essentially all (with protein IX being the sole exception) are transcribed from the major late transcription unit (MLTU). The MLTU is a 26-kb transcription unit that initiates at 16.5 map units (mu) and terminates in the region spanning map units 95–98 (**Fig. 1**).

Adenovirus has evolved a strategy of late gene expression that relies on a single transcription unit that uses expression of terminal exons to produce individual gene products. Through a combination of alternative splicing and alternative polyadenylation, in excess of 45 mRNA transcripts can be generated. Each transcript of the MLTU has a tripartite leader sequence (non-protein coding) spliced to a specific terminal exon (protein coding). The MLTU terminal exons are specified by RNA processing elements, a 5' splice accepter site, and one of five polyadenylation sites (L1–L5).

In the Ad MLTU, maintaining the integrity of the terminal exon is an important consideration when carrying out a gene-replacement strategy. It is also worth noting that the Ad MLTU has the capacity to express additional genes as distinct terminal exons. Therefore, expression of a nonviral gene late in infection or expression of multiple versions of a particular capsid protein (such as fiber) can be accomplished by insertion of an intact terminal exon. This strategy has evolved naturally in the subgroup F Ad, which contain two fiber genes (long and short). In the following protocol, we will describe procedures for a terminal exon replacement of Ad5 fiber with the tandem fiber arrangement of subgroup F Ad41.

2. Materials

1. 293 human embryonic kidney cells (or E1-complementing equivalent).
2. Standard molecular biology reagents for DNA manipulation:
   a. Restriction enzymes.
   b. Ligase.
   c. DNA phosphatase l.
3. *Escherichia coli* strains:
   a. Electroporation-competent BJ5183 cells from Stratagene (cat. no. 200154) (or equivalent rec$^+$ strain).
   b. Standard competent HB101 bacterial cells (or equivalent DH5α, DH10).
4. Proteinase K (20 mg/mL stock).
5. 1.0 M MgCl$_2$.
6. 1.0 M CaCl$_2$.
7. RNase A (10 mg/mL).
8. DNase I (5 mg/mL).
9. 10% Sarkosyl.
10. 0.5 M Ethylene diamine tetraacetic acid (EDTA).
11. 0.5 M Ethylene glycol tetraacetic acid (EGTA).
12. 4 M Ammonium acetate.
13. Isopropyl alcohol.
14. 70% Ethanol.

15. Plasmid strains:
    a. Donor vector (pAd70-100dlE3).
    b. Recipient vector (pvAdCiG).
    c. Intermediate shuttle vector pTARGET (Promega) or similar.
16. Standard reagents for tissue culture.
17. Tissue culture agarose (BioWhitaker).
18. 2X Minimum essential medium (MEM) (Gibco).
19. Calf serum (CS).
20. Lysis buffer (10 mL): 8.1 mL dH$_2$O, 0.5 mL 1.0 M Tris-HCl, pH 9.0, 0.4 mL 5% sodium salt of deoxycholic acid (DOC), 1.0 mL 100% ethanol (*see* **Note 4**).
21. Resuspension buffer (TNE): 100 mM NaCl, 10 mM Tris-HCl, pH 7.2, 1 mM EDTA.
22. Transfection TE: 10 mM Tris-HCl, pH 7.06, 1.0 mM EDTA.
23. 2X HeBS buffer: 140 mM NaCl, 1.5 mM Na$_2$HPO$_4$, 50 mM N-2-hydroxy-ethylpiperazine-N'-2-ethane sulfonic acid (HEPES).
24. 0.9% Agar preparation (100 mL total volume):
    a. Solution A: add 0.9 g tissue culture agar to 45 mL ddH$_2$O in an autoclavable bottle that can hold the appropriate final volume of agar overlay. Autoclave the solution for 30 min, then cool to 42–44°C (set up a water bath at this temperature).
    b. Solution B: 50 mL 2X MEM (Gibco/BRL), 2.5 mL CS, 1.0 mL 100X glutamine (200 mM in 0.85% NaCl; Gibco/BRL), 1.0 mL 100X antibiotic stock (pen/strep) (Gibco/BRL; 10,000 U/mL penicillin and 10,000 µg/mL streptomycin sulfate in 0.85% saline). Prepare this mixture in a sterile tube.
25. 100X Neutral red (3.333 g/L; Gibco/BRL).
26. GeneClean Spin Kits (QBiogene).
27. Calf intestinal alkaline phosphatase (CIP).
28. 15-mL Falcon tubes (cat. no. 2059) for chemical competent transformation.
29. Standard green fluorescent protein-expressing plasmid.
30. 10% Sodium dodecyl sulfate (SDS).
31. 2.5 M 2-Mercaptoethanol.
32. Tris-buffered phenol:chloroform.
33. Kanamycin (100X stock): 10 mg/mL.

3. Methods

The methods presented include the following:

1. Production of viral lysate stocks and production of viral DNA from the modified Hirt protocol.
2. Design of terminal exon for insertion and production of intermediate vector.
3. Manipulation of fiber domains.
4. Construction into a recombination plasmid.
5. Generation of terminal exon fiber virus.
6. Vector confirmation.

3.1. Small-Scale Primary Virus Lysate Stocks and Viral DNA

3.1.1. Viral Lysate Stock

The following protocol describes procedures that will generate a primary viral lysate stock.

1. Prepare a 60-mm dish of 293 cells, which will be 90% confluent in 36 h (5–10×10^5 cells per dish).
2. Prepare a dilution of virus that will yield approx 3–5 PFU/cell diluted into 1.0 mL of Dulbecco's modified Essential medium no Serum (DMEMnS) (no serum) (*see* **Note 1**).
3. Aspirate the medium from the monolayer and gently add the virus dilution to the dish (*see* **Note 2**). After 1 h, gently add 4 mL of fresh medium containing 2.5% CS. Place the dish in a 5% CO_2 incubator.
4. After 3–6 d (depending on the virus input) the cells will give evidence of cyto-pathic effect (CPE) (*see* **Note 3**). To harvest cells at full CPE, use a sterile plugged pipet to draw up the lysate and then wash the plate to dislodge any attached cells. Transfer to a sterile snap-cap or screw-top tube. Using dry ice or a freezer and a 37°C water bath, carry out three successive freeze–thaw cycles to release virus from cells. Centrifuge the lysate at 2000 rpm (1200g) and aliquot supernatant as 1-mL aliquots. Store at –75°C. Viral titer should be roughly 10^8 PFU/mL.

3.1.2. Modified Hirt Protocol for Viral DNA Preparation

This is a rapid and simple viral miniprep procedure. Depending on the virus and its growth characteristics, the procedure should yield 3–5 µg of viral DNA.

1. Set up a 10-cm dish of 293 cells such that they will be 90% confluent in 36 h.
2. Take 50 µL of a primary virus lysate and dilute in 1.0 mL of DMEM (ns). Add to cells and allow 1 h incubation at 37°C in a 5% CO_2 incubator. After 1 h add 8 mL DMEM containing 2.5% CS. Allow infection to proceed until full CPE is reached (approx 3 d, depending on virus input).
3. After cells have reached full CPE, harvest as previously described (**Subheading 3.1.1., step 4**) and transfer to a 15-mL sterile polypropylene tube.
4. Centrifuge cells at 300g for 10 min. Discard supernatant and resuspend cell pellet in 1.0 mL of PBS. Transfer to an Eppendorf tube and centrifuge 30 s. Discard supernatant.
5. Resuspend the cell pellet in 0.5 mL of lysis buffer and incubate for 1 h at room temperature (*see* **Note 4**).
6. Centrifuge in microfuge 5 min, and transfer supernatant to a fresh Eppendorf tube.
7. Add: 1.0 µL of 1.0 M $MgCl_2$, 1.0 µL of 1.0 M $CaCl_2$, 1.0 µL of RNase A (10 mg/mL), 1.0 µL of DNase I (5 mg/mL) to tube, mix gently, and incubate at 37°C for 30 min.
8. To each tube add: 25 µL 10% sarkosyl, 5 µL 0.5 M EDTA, 5 µL 0.5 M EGTA, and mix gently. Incubate at 70°C for 10 min.
9. Remove sample from 70°C bath and cool to room temperature. Add 2.5 µL of Proteinase K (20 mg/mL stock) and incubate 1–2 h at 37°C.

10. Extract with an equal volume phenol/chloroform (Tris-saturated). Microfuge 3 min and transfer supernatant to a fresh tube. Add one-tenth volume of 4 *M* ammonium acetate and add 750 µL of isopropyl alcohol. Chill at –70°C for 15 min.
11. Centrifuge the sample for 15 min, discard the supernatant, and wash the DNA pellet with 1.0 mL of 70% ethanol. Pour off the supernatant and air-dry the DNA pellet.
12. Resuspend the pellet in 30 µL of TE.
13. Run 3 µL on a minigel to determine yield.

At this point you will have generated enough viral DNA for restriction digestions or for polymerase chain reaction (PCR) amplification of your region of interest. This procedure can be scaled up to produce more DNA if desired.

3.2. Identification of the Target Terminal Exon and PCR Amplification

Human adenoviruses have a well-conserved genome organization, and the terminal exons of the MLTU are well defined (**Fig. 1**). Using readily available sequence databases, it is a straightforward matter to identify the location of the target capsid protein in your serotype of interest. We will go through a strategy designed for generating an Ad5-based vector that replaces the Ad5 fiber with the tandem fibers of Ad41. Fiber is coded within the L5 terminal exon (Hexon in L3, Penton in L2). We use DNA Strider as a simple software package for basic sequence analysis.

1. Knowing the relative location of the target gene, it is useful to identify the proper open reading frame (ORF) for the target capsid protein. Identify the translation initiation codon ATG. The termination codon is revealed when doing a primary translation of the DNA in frame with this ATG. It is always important to confirm several times that you are targeting the right sequence. An old carpenter's rule is: "measure twice and cut once."
2. The procedure that we describe is a terminal exon replacement of fiber. To accomplish this we must identify the essential RNA-processing elements that define the fiber terminal exon; these elements include the fiber splice acceptor and the fiber poly A site (L5).
   a. The poly A site is identifiable by the highly conserved AATAAA hexameric sequence that is immediately 3' to the ORF termination site. The poly A cleavage site is usually 15–20 nucleotides 3' to the AATAAA, and an additional 20 nucleotides 3' to the cleavage site usually contributes to cleavage efficiency (**Fig. 2**). PCR primers should be created such that the downstream domains are included in the amplified terminal exon domain.
   b. The 5' splice acceptor (SA) site does not have as highly a conserved sequence motif, but can be easily identified, and is located 5' to the ATG of the fiber ORF. The 5' SA site has a highly conserved AG:A trinucleotide where cleavage occurs between the G:A residues. Immediately upstream of the conserved trinucleotide is a polypyrimidine tract and a branch point A. (In a remarkable

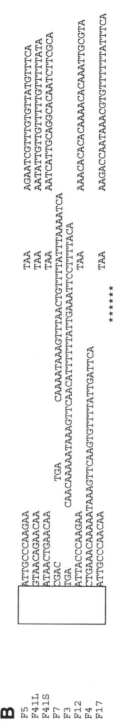

Fig. 2. Sequence conservation of fiber splicing and 3'-processing element sites. (A) Fiber DNA sequences upstream of the corresponding translation initiation codon (AUG) were aligned using ClustalW. Conserved residues are denoted with an asterisk. Sequences corresponding to pre-mRNA splicing elements are indicated as follows: splice acceptor (boxed), polypyrimidine tract (dashed line), putative branch site (solid line). (B) 3'-Fiber sequences were aligned manually at the consensus hexanucleotide AATAAA (*). Downstream elements include sequences corresponding to U/GU-rich regions in fiber pre-mRNA and putative adenine cleavage sites. Fiber translation stop codons are underlined.

example of efficiency, in many Ad serotypes, the fiber splice acceptor sequence is coincident with the ATG translation initiation codon, and the polyadenylation AATAAA is coincident with the translation termination codon. PCR primers should be created such that the polypyrimidine tract and the branch point A are included in the amplified terminal exon domain.)

3. Splice acceptor 5' PCR primer oligo: the 5' oligo is designed to complement the specific sequence that begins at the branchpoint A of the 5' splice acceptor. In the case of the fiber 41 tandem genes, the sequence upstream of the initiation ATG is: CCTTTTTACCCTGACCCACGATCTTCATCTTGCAG*ATG*. We selected the sequence immediately upstream of the branchpoint A for the Ad41 short 5' splice acceptor: CCTTTTTACCCTGACCC. For cloning purposes, we add sequences that will allow cloning into both the expression cassette and into the viral recombination cassette. In our case we have added sequences for *Bam H*I and *Pac*I sites and a few buffer nucleotides to protect the ends of the PCR products: cccggatccttaattaa. The 5' splice acceptor primer for F41 tandem is 5'-cccggatccttaattaa (CCTTTTTACCCTGACCC) and corresponds to the coding strand of fiber.

4. Poly A site 3' PCR primer oligo: based on the position of the AATAAA hexanucleotide located in the coding strand of viral DNA and in order to include the majority of sequence that influences polyadenylation, simply count 50 nucleotides downstream (3') of the AATAAA. This should give sufficient sequence downstream of the AATAAA to include the authentic mRNA cleavage site (at approx 20 nts) as well as sequence that contributes to the efficiency of the 3' RNA cleavage event. In this instance the 3' primer would correspond to the reverse complement of the coding strand. In the case of Ad41, **AATAAA**ATA TTGTTGTTTTTGTTTTTAT**A***ACTTTATTGATCATTTTACAGAATT* represents the coding sequence of interest. The sequence that we select for our PCR primer is indicated by bold/italicized letters. The reverse complement of this sequence 5'-AATTCTGTAAAATGATCAATAAAG corresponds to the primer necessary for amplification from the coding strand. To the 5'-end of this primer, restriction enzyme sites should be added for the purpose of cloning. We have added a BamHI linker (GGATCC) as well as buffer sequence to facilitate efficient digestion. The final 3' primer is cccggatcc(AATTCTGTAAAATGA TCAATAAAG).

5. PCR amplification from miniprep viral DNA: amplification from viral miniprep DNA is straightforward and should be accomplished using standard procedures. For routine 50-µL amplifications, 30 ng of viral DNA is sufficient template.

3.3. Fiber Expression Cassette

As a matter of cloning convenience, we first build the fiber gene into an intermediate plasmid (any basic plasmid vector will work). We use a pTarget that has a modified poly linker to accommodate our particular cloning strategy. This allows us to express the target protein in the absence of virus construction. It also allows simple confirmation of the PCR-amplified region by DNA

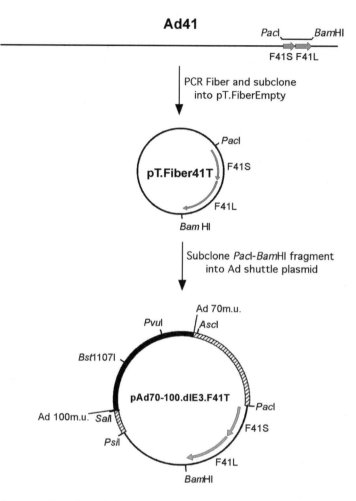

Fig. 3. Schematic drawing of the steps involved in production of pAd70100dl.F41T shuttle plasmid. From viral DNA, the target fiber region is PCR amplified, subcloned into an intermediate vector pTarget, and then moved into the virus recombination shuttle vector by PacI/Bam HI cloning. Digestion of this plasmid with *Sal*I and *Asc*I releases the Ad 70–100 region from the plasmid backbone.

sequencing and provides a vector that is easy to handle if we are interested in carrying out sequence manipulations of the fiber of interest.

3.4. Viral Cloning Vector pAd5 70-100

We use the plasmid pAd70-100dlE3 (**Fig. 3**) for carrying out virus construc-
tions. This plasmid shuttle system is compatible with recombination strategies used in bacteria as well as those used in mammalian cell lines. The essential

features of the shuttle vector are highlighted in **Fig. 3**. This plasmid contains a modified Ad5 fiber terminal exon. The vector contains a unique *Pac*I restriction site upstream of the fiber splice acceptor such that all of the accessory splicing sequences have been preserved. A *Bam* HI restriction site has been inserted downstream of the L5 poly A site, but upstream of the E4 poly A site E4 transcription unit, which is transcribed from the opposite DNA strand. The plasmid contains several kilobases of Ad5 sequence flanking both the 5'- and 3'-ends of the fiber terminal sequence. The Ad flanking sequence provides a generous region for overlap recombination. The shuttle vector extends to Ad m.u. 100 at the *Sal*I site, including the right-end viral origin of replication.

3.5. Making a Recombinant Adenovirus Containing a New Terminal Exon in the MLTU

Two methods are described for generating a recombinant virus; the first uses recombination in bacterial cells *(4,5)*. The second procedure is direct recombination in complementing 293 cells *(6)*. The principles behind both strategies are similar. The general strategy relies on cotransfection of a linearized pAdCiG viral vector (recipient) with a linearized pAd70-100 F(X) modified plasmid (donor) (**Fig. 4**). In bacteria, through double-recombination between regions of homology and kanamycin selection, a recombinant plasmid vector is obtained that has the new terminal exon (F41T) in place of the original fiber. Similarly, when producing the virus directly in 293 cells, a single overlap recombination occurs between the viral vector and the shuttle plasmid. When a successful recombination has occurred, viable virus is recovered. Recombination in bacteria has several benefits with respect to speed and greater flexibility in plasmid management.

3.5.1. Construction of a Recombinant Viral Plasmid in BJ5183 Bacterial Cells

The parent plasmid (also referred to as the recipient plasmid) we use for recombination is pAdCiG (**Fig. 4**), which contains the full Ad5 E1/E3-deleted vector genome and expresses a bicistronic, cytomegalovirus-driven CATiresGFP element in the E1 region. The MLTU contains a PacI/BamHI-modified fiber 5 terminal exon. The plasmid backbone of pAdCiG contains a kanamycin-resistance cassette as well as the Col E1 origin of replication. The plasmid backbone can be released from the Ad sequences by *Swa*I digestion. We purchase electroporation-competent BJ5183 (Stratagene). Alternatively, a procedure for chemically competent transformation is also presented.

1. Digest 5 µg pAdCiG with *Psi*I and dephosphorylate with CIP. This digest releases fragments from the right-most end of the Ad genome, including the 3' half of the fiber gene (**Fig. 4**).

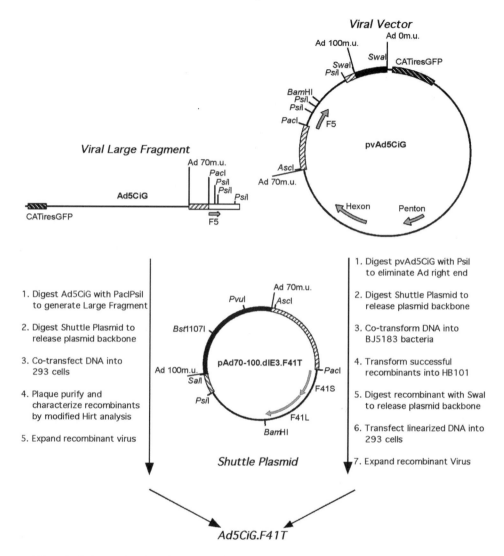

Fig. 4. Schematic drawing depicting conversion of fiber shuttle plasmid into a recombinant adenovirus.

2. Prepare the donor plasmid by digestion of 5 µg of pAd70-100.dlE3.FiberX with *Asc*I and *Sal*I. This digest releases the fiber gene with sufficient flanking sequence to serve as homology arms for recombination.
3. Purification of the DNAs for recombination is not required, but it can reduce background artifacts (if gel purification is not performed, then a phenol: chloroform extraction and ethanol precipitation of the restriction digests is nec-

essary). Gel electrophoresis of digested products in a 0.8% low-melt agarose gel provides good separation of the DNA fragments of interest (~33 kb for the vector and ~8.5 kb for the shuttle). We use GeneClean Spin Kits for routine DNA purification from agarose gels. DNA must be eluted from the spin column in ultra-pure dH_2O to prevent ionic interference during electroporation.

4. Combine approx 50–100 ng recipient and 100–300 ng donor fragments in a sterile Eppendorf tube for electroporation or a 15-mL Falcon 2059 for chemical competent transformation.

5. For electroporation:
   a. Add 25–30 µL competent cells to prechilled Eppendorf tubes on ice. Add the DNA mixture to competent cells and transfer the mixture to a chilled electroporation cuvet. (Include all appropriate control samples for background and transformation efficiency.)
   b. Electroporation conditions are 200 Ohms, 2.5 kV, 25 µF. Pulse the sample one time and add 1 mL prewarmed sterile SOC asis Luria broth (LB). Resuspend by pipetting up and down.
   c. Transfer the mixture to a 15-mL Falcon 2059 tube and shake at 250 rpm for 15 min at 30°C.

6. For chemical-competent transformation:
   a. Add approx 50 µL of $CaCl_2$-competent cells to the 2059 tube and heat-shock at 45°C for 45 s.
   b. Add 1 mL prewarmed SOC or LB and incubate at 30°C at 160 rpm for 1 h.

7. Plate out the entire 1-mL suspension onto three LB plates containing kanamycin (100 µg/mL): 750, 200, and 50 µL. Incubate plates for more than 18 h at 30°C.

8. Pick 12 colonies into 2 mL LB + kanamycin and incubate overnight at 30°C, 160 rpm.

9. Prepare miniprep DNA from 1.5 mL of the overnight culture. Include a phenol:chloroform extraction before ethanol precipitation.

10. Digest an aliquot of miniprep DNA with enzymes appropriate to diagnose a successful recombination. In addition, PCR can be performed on miniprep DNA using primers specific to the novel fiber being inserted. The parent plasmid pAdCiG should be used as a control for all miniprep digests and PCR.

11. Transform miniprep DNA from successful recombinants into standard competent bacterial cells, such as DH10, DH5α, or HB101 cells. Several colonies should be picked, grown for miniprep DNA, isolated, and verified by restriction digest and PCR.

3.5.2. Transfection of Successful Recombinants in 293 Cells (see **Note 5**)

1. Digest approx 5 µg recombinant DNA with SwaI to release the kanamycin-resistance cassette. Run a small aliquot of the reaction on a minigel to verify digestion.

2. Increase the reaction volume to 200 µL with TNE and extract with an equal volume of phenol:chloroform. Transfer aqueous phase to a fresh tube.

3. Precipitate DNA with 2 vol ethanol. Microfuge for 10 min. Decant supernatant and wash 1X with 70% ethanol. Air-dry pellet in laminar flow hood.

4. Resuspend DNA and transfect into 293 cells (as described in **Subheading 3.6.2.**). For transfections, we include a separate condition containing a standard GFP-expressing plasmid.
5. GFP from transfected cells should be evident 24–36 h posttransfection. Five to 7 d later, harvest cells in 15-mL tubes and freeze–thaw three times to release virions. Reinfect 293 cells using an aliquot of the lysate for amplification of virus (*see* **Note 6**). Successful virus production will be evident by the presence of GFP-expressing cells.
6. As described previously, modified Hirt assays can be done from 10-cm plate infections to verify the integrity of recombinant DNA by restriction digest and/or PCR.
7. If the recombinant is verified by modified Hirt analysis, then the virus can be amplified for large-scale virus production using standard techniques.

3.6. Recombination in 293 Cells: Fiber Shuttle Plasmid and Viral Large Fragment

The production of novel fiber-modified vectors can be carried out by homologous recombination in 293 cells by transfecting a viral large fragment DNA corresponding to the left end of the Ad genome (**Fig. 4**) with a donor plasmid bearing the fiber-containing right end of the Ad genome *(7,8)*. Viral large fragment can be obtained from plasmid backbones such a pvAdCiG or from viral DNA. A protocol for purifying viral DNA from a virus stock (10^{12} particles) is described.

3.6.1. Small-Scale Purification of Viral Vector DNA

1. Increase volume of virus to 450 µL by adding 350 µL of resuspension buffer.
2. Add one-tenth volume of 10% SDS (45 µL). Vortex and incubate at room temperature for 30 min.
3. Add one-tenth volume 2.5 *M* 2-mercaptoethanol (50 µL). Vortex and incubate at room temperature for 30 min.
4. Add 15 µL Proteinase K (20 mg/mL). Vortex and incubate 1–2 h at 37°C.
5. Extract 2X with an equal volume of Tris-buffered phenol:chloroform. Vortex briefly and microfuge for 1 min. Transfer aqueous phase to new tube.
6. Precipitate DNA by adding 20 µL of 5.0 *M* NaCl and 2 vol of 100% ethanol.
7. Microfuge sample for 20 min. Pour off supernatant and wash 1X with 70% ethanol. Air-dry pellet 5–10 min.
8. Resuspend DNA in 30 µL TE.
9. The OD_{260} of a 2 µL aliquot can be used to determine DNA yields (*see* **Note 7**).

3.6.2. DNA Digestion and Transfection

3.6.2.1. DNA DIGESTION

1. Digest 10 µg of vector DNA with *Psi*I and *Pac*I to generate the left-end large fragment.

2. Digest 10 μg of plasmid DNA pAd70-100.dlE3.FiberX with *Pvu*I and *Bst* 1107I to release plasmid backbone sequences from right-end Ad DNA (*see* **Note 8**).
3. Run a small aliquot of each reaction on an agarose minigel to verify digestion.
4. Purify DNA by phenol:chloroform extraction, followed by ethanol precipitation and a 70% ethanol wash. Allow pellet to air-dry in a sterile laminar flow hood.

3.6.2.2. TRANSFECTION

The following steps should be carried out using standard sterile tissue culture techniques. DNA transfections are carried out on 60-mm dishes of 70–80% confluent 293 monolayers.

1. Resuspend DNA in 15 μL of sterile transfection TE.
2. Estimate DNA concentrations by analyzing 2-μL aliquots on an agarose minigel alongside a DNA sample of known concentration.
3. Set up two tubes for transfection: (1) vector only, and (2) vector + fiber. (A "vector-only" control is included to determine the presence of background virus from incompletely digested vector DNA.)
4. To each tube add 2 μg of *Psi*I/*Pac*I-digested vector DNA. To tube 2, add 5 μg of *Pvu*I/*Bst* 1107I-digested pAd70-100.dlE3.FiberX plasmid. To tube 1, add 5 μg of any nonviral plasmid (~10–13 kb) as a transfection carrier control. Increase volume in each tube to 125 μL with sterile dH$_2$O.
5. Add 125 μL 0.5 *M* CaCl$_2$ to each DNA sample and mix by pipetting.
6. Prepare two 5-mL polystyrene snap-cap tubes, each containing 250 μL of prewarmed (37°C) 2X HeBS buffer.
7. Add the 250 μL DNA/CaCl$_2$ solution to 250 μL 2X HeBS in a dropwise fashion over a low-to-medium vortex.
8. Incubate sample for 1 min. Add solution to a 293 monolayer. The following day, remove medium, wash cells 1X with fresh DME (5% CS), and add 5 mL fresh medium.
9. At 7–10 d posttransfection cytopathic effect should be clearly evident.
10. At full CPE, harvest cells (described in **Subheading 3.1.**). The virus generated in this manner is a possible mixture of the desired recombinant and background virus. The characterization of modified Hirt DNA should indicate the degree of successful recombination (*see* **Note 9**).

3.6.3. Plaquing Adenovirus on 293 Monolayers

1. Prepare 293 monolayers on 60-mm dishes such that they will be 70–80% confluent in 36 h. For each virus to be plaqued, prepare at least three dishes. If you are titrating virus by the traditional plaque assay, set up duplicate plates for each dilution.
2. Starting with a virus lysate of unknown titer estimate a titer of 10^8 PFU/mL, prepare serial dilutions of your virus at 10^6, 10^4, and 10^3 PFU/mL final concentration of virus (in DME). Use 1, 10, 25, and 100 μL of the 10^3 dilution to a final volume of 0.2 mL.

3. Gently aspirate medium from the 293 monolayer, and gently add the virus dilution. Rock and incubate for 1 h.
4. Add 5.0 mL of medium to the plate and let incubate overnight. The following morning, prepare the agar overlay (8 mL/dish).
5. When you are ready to overlay your infected 293 monolayers, bring your dishes into the hood, and carefully aspirate medium from the plates.
6. Add solution B to solution A (**Heading 2.**, **item 24**) and mix without generating too many air bubbles. The agar will set shortly so you must work efficiently at this point.
7. Using a sterile plugged 25-mL pipet, overlay 7 mL of agar onto the dish (*see* **Note 10**). Allow the agar to cool for at least 30 min at room temperature.
8. Place in CO_2 incubator at 37°C.
9. Plaques should be visible "by eye" in 3–5 d without staining.
10. To better visualize the plaques, a second overlay containing neutral red can be used to highlight areas of uninfected cells (revealing an adenoviral plaque as a clearing zone). Use the same formula as presented above (**step 6**), but use 43 mL H_2O (for solution A) rather than 45 mL (per 100 mL of agarose overlay) and add 2.0 mL 100X neutral red. Overlay each plate with 3 mL of the final mixture.
11. Depending on the purpose of your plaque assay, you can count your plaques on each plate and calculate the PFU/mL of the original stock.

3.6.4. Pick Individual Plaques for Further Characterization or Use

1. Set up a dozen small snap-cap tubes for each virus.
2. Add 2 mL of DMEMns to each tube.
3. Use a cotton-plugged sterile Pasteur pipet to draw out the agar plug directly over your plaque (*see* **Note 11**).
4. Transfer the plug to the tube by drawing the medium into your pipet and washing the plug into the tube.
5. You can store the virus at –70°C until you are ready to grow the virus.
6. When you are going to process the plaque isolate, freeze–thaw the mixture an additional two times. Use 0.5 mL of the virus-containing medium to infect a 60-mm dish as previously described. Depending on the plaque size and the amount of virus in the plug, signs of CPE will start to become obvious at or after 4 d. Wait until full CPE has occurred before harvesting; this should be between 5 and 8 d.

3.7. Insertion of Novel Terminal Exons Into the MLTU

The strategies presented in this chapter have illustrated how a specific fiber-replacement strategy can be executed to alter the fiber content of Ad5. The protocols are straightforward applications of generally available recombinant technologies. Using the general strategy as outlined, there is a great deal of flexibility in the type of terminal exon replacement that can be carried out. Manipulations of fiber domains, insertion of unique gene products that will be expressed specifically during the late stage of infection, or expression of multiple versions of a specific capsid protein are possible. In our experience, we have found that not all terminal exons are recognized with the same efficiency as the Ad5 fiber. As an example, expression of the F7 terminal exon was more than sufficient to make a viable virus, but the amount of steady-state processed mRNA was more than 2 logs less than that generated by the fiber 41S terminal exon when co-expressed from the same virus *(8)*. Importantly, the relatively low use of the F7 terminal exon recognition is an inherent feature of the F7 fiber and occurs in wild-type Ad7A infections as well. Much as promoters vary in a sequence-dependent manner, terminal exons may also vary in terms of their efficiency of use.

4. Notes

1. If you are getting virus from the American Type Culture Collection, they will indicate a viral titer. If you do not know the titer, one can estimate that most stocks are 10^7–10^8 PFU/mL or higher.
2. Handle 293 monolayers carefully; they are fragile.
3. Cells will begin to round up and detach from the dish. Harvest the plate when full CPE has occurred, i.e., all the cells are demonstrating this altered phenotype. Full CPE is often accompanied by acidification of the medium (medium turns orange or yellow).
4. The lysis buffer should be made just before use.
5. In our hands, miniprep DNA from HB101 that has been RNaseA-treated and phenol:chloroform-extracted is sufficient for transfection into 293 cells. Alternatively, cultures can be amplified and the DNA purified using any larger scale procedure that provides endotoxin-free DNA.
6. For generating recombinants expressing fibers that bind poorly to 293 cells, we use high volumes of lysate on a small number of cells (i.e., 2 mL of lysate into one well of a 24-well plate). This allows for a faster expansion of growth-defective viruses.
7. We routinely recover 20–50 μg of viral DNA depending on the particle/mL concentration of the starting prep.
8. As described for recombination in bacteria, gel-purification of both the donor and recipient vectors can be used to decrease background.
9. The burden of using viral DNA as the backbone for generating a viral recombinant is the need to plaque-purify your recombinant virus. We recommend two rounds of plaque purification.
10. 293 Monolayers can be very easily displaced from the dish; gently start overlaying by dispensing the agar down the side wall of the dish. After the entire dish is covered, add the remainder of the overlay over the residual DME that usually localizes to one spot of the dish. This will prevent a soft spot from occurring in the overlay.
11. We use a pipet aid to gently draw the plug into the pipet.

References

1. Bergelson, J. M. (1999) Receptors mediating adenovirus attachment and internalization. *Biochem. Pharmacol.* **57,** 975–979.
2. Wickham, T. J., Mathias, P., Cheresh, D. A., and Nemerow, G. R. (1993) Integrins alpha v beta 3 and alpha v beta 5 promote adenovirus internalization but not virus attachment. *Cell* **73,** 309–319.
3. Nicol, C. G., Graham, D., Miller, D. W., et al. (2004) Effect of adenovirus serotype 5 fiber and penton modifications on in vivo tropism in rats. *Mol. Ther.* **10,** 344–354.
4. Chartier, C., Degryse, E., Gantzer, M., Dieterle, A., Pavirani, A., and Mehtali, M. (1996) Efficient generation of recombinant adenovirus vectors by homologous recombination in Escherichia coli. *J. Virol.* **70,** 4805–4810.
5. McVey, D., Zuber, M., Ettyreddy, D., Brough, D. E., and Kovesdi, I. (2002) Rapid construction of adenoviral vectors by lambda phage genetics. *J. Virol.* **76,** 3670–3677.
6. Haj-Ahmad, Y. and Graham, F. L. (1986) Development of a helper-independent human adenovirus vector and its use in the transfer of the herpes simplex virus thymidine kinase gene. *J. Virol.* **57,** 267–274.
7. Gall, J., Kass-Eisler, A., Leinwand, L., and Falck-Pedersen, E. (1996) Adenovirus type 5 and 7 capsid chimera: fiber replacement alters receptor tropism without affecting primary immune neutralization epitopes. *J. Virol.* **70,** 2116–2123.
8. Schoggins, J. W., Gall, J. G., and Falck-Pedersen, E. (2003) Subgroup B and F fiber chimeras eliminate normal adenovirus type 5 vector transduction in vitro and in vivo. *J. Virol.* **77,** 1039–1048.

9

Temperature-Sensitive Replication-Competent Adenovirus shRNA Vectors to Study Cellular Genes in Virus-Induced Apoptosis

Thirugnana Subramanian and Govindaswamy Chinnadurai

Summary

The use of shRNA for knockdown of gene expression is a powerful method. In addition to transient transfection of RNA oligonucleotides, various DNA-based vectors that express short hairpin RNAs have been successfully used for efficient depletion of gene products. Replication-defective retrovirus and adenovirus (Ad) vectors have also gained wide usage. The extension of shRNA technology to replication-competent Ad would be desirable to investigate the role of various cellular genes in Ad replication. This approach is hampered because the effect of shRNA is neutralized by the Ad VA-RNA that is expressed at late stages after infection, and the infected cells are killed prior to significant depletion of some long-lived target gene products. We have constructed replication-competent Ad vectors for the depletion of the pro-apoptotic proteins BAX and BAK. We have modified a replication-defective Ad multivalent shRNA expression vector developed by Welgen, Inc. In our vector design, the multivalent shRNA expression cassette is contained in the E1B region. Additionally, we have incorporated a temperature-sensitive mutation in the viral *DBP* gene (*ts*125). The use of this vector has resulted in efficient depletion of critical cellular apoptotic modulators, BAX and BAK. This vector may be useful to study the role of various cellular genes in Ad-induced apoptosis and viral replication.

Key Words: Adenovirus; viral vector; shRNA; RNAi; BAX; BAK; H5ts125.

1. Introduction

RNA interference (RNAi) is a powerful tool for the silencing of gene expression. It is a sequence-specific gene knockdown process that occurs at the posttranscriptional level *(1,2)*. RNA interference can be induced in cells by transient transfection of double-stranded RNA oligonucleotides or DNA-based plasmid vectors that express target gene-specific short hairpin RNAs (shRNAs). Several replication-defective retrovirus and adenovirus (Ad) vec-

From: *Methods in Molecular Medicine, Vol. 130:*
Adenovirus Methods and Protocols, Second Edition, vol. 1:
Adenoviruses Ad Vectors, Quantitation, and Animal Models
Edited by: W. S. M. Wold and A. E. Tollefson © Humana Press Inc., Totowa, NJ

tors have also been successfully used to express shRNAs. Many laboratories have constructed Ad vectors to express shRNA under the transcriptional control of RNA polymerase III promoters, such as the human H1, human U6, and mouse U6 promoters. Zhao et al. *(3)* have reported efficient suppression of p53 under the control of the human H1 promoter in a replication-defective (E1-minus) Ad vector. Davidson and colleagues have described an Ad vector that expresses shRNA under the control of a mutant CMV promoter *(4)*. Several biotechnology companies also market different Ad-based shRNA vectors. Welgen Inc. (Worcester, MA) markets multivalent shRNA Ad vectors that could be used to simultaneously target multiple genes. Others include Invitrogen (Carlsbad, CA), GenScript (Piscataway, NJ), and BD Biosciences (San Jose, CA).

All currently available Ad vectors are replication-deficient as a result of the deletion of the E1A and E1B regions. However, they potentially could be used as conditional-replication vectors in cells that constitutively express E1A and E1B regions to investigate the role of different cellular genes in virus-induced apoptosis and replication. However, these vectors appear to impose two limitations. First, the effect of certain long-lived proteins could not be satisfactorily investigated because the infected cells would lyse between 30 and 40 h after infection. Second, the expression of VA-RNA during late stages of productive viral replication would inhibit the activity of shRNAs *(5)*. In this chapter we describe the construction of replication-competent adenoviral vectors to deplete human BAX and BAK proteins. Our vector designs incorporate a temperature-sensitive mutation, ts125 *(6,7)*, in the viral *DBP* gene. This design facilitates delay of the viral replication under nonpermissive temperature (39.5°C) for an extended period of time to achieve satisfactory depletion of the target gene products. Although we have designed the present vectors to study the role of cellular genes in Ad-induced apoptosis, the design could be readily modified to investigate the role of various cellular genes in viral replication as well.

2. Materials

1. pQuiet-4 vector (purchased from Welgen, Inc.).
2. pLendΔE1BCMV (constructed in our Lab by introducing E1A coding sequences into pLendCMV; *see also* **ref. 8**).
3. pacAd5 9.2-100 (gift from Beverley Davidson; **ref. 4**).
4. H5ts125 DNA: the viral DNA was prepared from a stock of H5ts125 mutant purified by banding in CsCl *(6,7)*.
5. Restriction enzymes and T4 DNA ligase.

6. Oligonucleotides: use salt-free unpurified oligonucleotides (purchased from Operon Biotechnologies, Inc., Huntsville, AL).
7. TE buffer: 10 mM Tris-HCl, pH 8.1, 1 mM EDTA.
8. 10X Annealing buffer: 500 mM Tris-HCl, pH 7.6, 20 mM MgCl$_2$.
9. *Escherichia coli* competent cells: BJ5183, DH5α and HB101.
10. Agarose and polyacrylamide gel electrophoreses reagents and apparatus.
11. Western blot transfer unit, buffers, antibodies (anti-Bax and anti-Bak rabbit antiserum [Upstate, Charlottesville, VA] and horseradish peroxidase (HRP)-conjugated goat anti-rabbit IgG [Santa Cruz Biotechnology, Santa Cruz, CA]) and chemiluminescent detection systems (Roche Applied Sciences, Indianapolis, IN).
12. 2X HeBS: dissolve 8.182 g of NaCl, 5.958 g of N-2-hydoxyethylpiperazine-N'-2-ethanesulfonic acid (HEPES), and 107 mg of NaHPO$_4$ in 400 mL of water and adjust the pH to 7.10 $\pm$ 0.05. Make up the volume to 500 mL, filter-sterilize, and store at -20°C.
13. 2 M CaCl$_2$: dissolve 22.2 g of CaCl$_2$·7H$_2$O in 100 mL of water, filter-sterilize, and store at -20°C.

3. Methods

3.1. Construction of pLendΔE1BQ4

In order to express the shRNA with two different target sequences of a single gene, we cloned the multivalent shRNA expressing cassette (Welgen, Inc.) into our Ad left-end plasmid that has a deletion of the E1B region (pLendΔE1B) between two engineered unique restriction sites for *Spe*I and *Asc*I (pLendΔE1B-S/A). This plasmid was used for the construction of shRNA expression plasmids according to the following protocol. (A similar strategy could also be used for mobilization of the multivalent shRNA cassette to other nonessential regions such as the E3 region.)

3.1.1. Cloning of Bidirectional Pol III Promoter Cassette Into pLendΔE1B-S/A

1. Digest 5 μg of pLendΔE1B-S/A with *Spe*I and *Asc*I and purify the vector by agarose gel electrophoresis.
2. Digest 5 μg of pQuiet4 vector with *Spe*I and *Asc*I and purify the fragment containing a bidirectional human U6 promoter at positions 1 and 2 and a bidirectional hybrid human U6 and H1 promoter at positions 3 and 4 with restriction sites for the insertion of target sequences (**Fig. 1A**).
3. Ligate the purified fragment and the vector.
4. Transform HB101 competent cells, screen the ampicillin-resistant colonies by digestion of mini preparation of DNA with *Asc*I.
5. Perform DNA sequence analysis of the clone to confirm and name it as pLendΔE1BQ4 (**Fig. 1B**). This vector is available from the authors upon request.

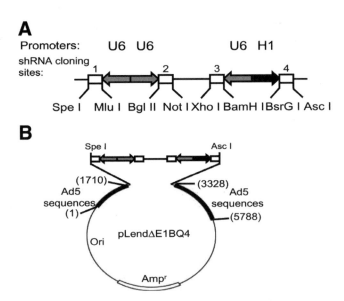

Fig. 1. Map of multivalent shRNA cassette of pQuiet4 (**A**). The four cloning sites for the insertion of different target sequences and the bidirectional promoters are shown at the top. The first bidirectional promoter consists of two human U6 promoters fused together in opposite orientation and the second is a hybrid bidirectional promoter with human U6 and H1 fused in opposite orientation. The cloning sites are indicated at the bottom. Map of pLendΔE1BQ4 (**B**). The E1B region encompassing nucleotides 1710 and 3328 is replaced with the multivalent cloning cassette of the pQuiet-4 vector. The numbers in parentheses indicate the nucleotide position of Ad5 sequence.

3.2. Cloning of Human BAX Target Sequences at Sites 1 and 2 of pLendΔE1BQ4

We selected two target sequences for human *BAX* gene using previous strategies *(9)*. One such target sequence is from nucleotide positions 222 to 240 and the other from 156 to 174 of the BAX ORF. The target sequences for genes of interest could be selected on the basis of various algorithms currently available.

3.2.1. Designing Oligonucleotides for the First Target Sequence at Site 1 of pLendΔE1BQ4 Vector

The oligonucleotides are designed according to a protocol provided by Welgen, Inc. For each target sequence, make two complementary strands with appropriate restriction overhangs at the ends. The top strand is engineered with the overhang sequence for *Mlu*I followed by 19 bases of BAX target sequence (sense strand, indicated with brackets), 9 bases for hairpin loop, 19 bases of antisense nucleotide, five T, and at the end one A for recovering *Spe*I site and the bottom strand with *Spe*I overhang followed by complementary sequence of top strand and at the end one A for recovering *Mlu*I site.

Site 1, Top Strand: 5'-CGCGT[GGAGCTGCAGAGGATGATT] TTCAAGAGA AATCATCCTCTGCAGCTCC TTTTT A-3'
Site 1, Bottom Strand: 5'-CTAGT AAAAA GGAGCTGCAGAGGATGATT TCTCTTGAA AATCATCCTCTGCAGCTCC A-3'

Anneal both oligonucleotides and ligate to the *Mlu*I and *Spe*I digested pLendΔE1BQ4 vector as mentioned in **Subheading 3.1.1.** Name the plasmid with the first target sequence of human BAX at site 1 of pLendΔE1BQ4 as pLendΔE1BQ4-1BAX.

3.2.2. Designing Oligonucleotides for the Second Target Sequence at Site 2 of pLendΔE1BQ4-1BAX Vector

Design oligonucleotides as shown for the first target sequence with the overhang modification. The second site of pLendΔE1BQ4-1BAX vector has *Bgl*II and *Not*I restriction sites for cloning and according to the restriction sites synthesize the following oligonucleotides for site 2 of pLendΔE1BQ4-1BAX vector.

Site 2, Top Strand: 5'-GATCT GGATGCGTCCACCAAGAAG TTC AAGAGA CTTCTTGGTGGACGCATCC TTTTT GC-3'
Site 2, Bottom Strand: 5'-GGCCGC AAAAA GGATGCGTCCACCAAGAAG TCTCTTGAA CTTCTTGGTGGAGCGATCC A-3'

Anneal the above oligonucleotides and ligate to the *Bgl* II and *Not* I digested pLendΔE1BQ4-1BAX vector. Name the plasmid with both target sequences of human BAX at sites 1 and 2 of pLendΔE1BQ4 as pLendΔE1BQ4-1,2BAX.

3.3. Cloning of Human BAK Target Sequences at Sites 3 and 4 of pLendΔE1BQ4

The authors have selected two target sequences of human BAK gene that were previously reported to inhibit BAK expression *(10,11)*. These target sequences were cloned into pLendΔE1BQ4 vector at sites 3 and 4. The target sequences are from positions 256 to 276 and from 324 to 342 of human BAK coding sequence numbering from the initiation codon. According to the restriction sites availability, *Xho*I and *Bam*HI overhangs for site 3 and *Bsr*GI and *Asc*I overhangs for site 4 were engineered in the oligonucleotides. The oligonucleotides used for cloning human BAK shRNA expression vector were as follows:

Site 3, Top Strand: 5'-TCGAG AACCGACGCTATGACTCAGAG TTCAAGA GACTCTGAGTCATAGCGTCGGTT TTTTT G-3'
Site 3, Bottom Strand: 5'-GATCC AAAAA AACCGACGCTATGACTCAGAG TCTCTTGAA CTCTGAGTCATAGCGTCGGTT C-3'

Site 4, Top Strand: 5'-GTACA TGAGTACTTCACCAAGATT TTCAAGAGA
AATCTT GGTGAAGTACTCA TTTTT GG-3'
Site 4, Bottom Strand: 5'-CGCGCC AAAAATGAGTACTTCACCAAGATT
TCTCTTGAA AATCTTGGTGAAGTACTCA T-3'

The authors recommend cloning into sites 3 and 4 sequentially. Although
the multivalent vector has the potential to target four different genes simultaneously, we have only examined the effect on two different genes.

3.4. Introducing ts125 Mutation Into pacAd5 9.2-100

We have used in vivo homologous recombination in *E. coli (12)* to introduce the H5ts125 mutation into pacAd5 9.2-100 according to the following
protocol.

1. Digest pacAd5 9.2-100 with BstZ17 I and gel-purify the vector and the attached
   Ad5 sequences (there are two sites for *BstZ17* I in the Ad5 genome; at nucleotide
   positions 5764 and 29010 and no site in the vector).
2. Take OD and calculate the amount of vector.
3. Dilute the digested vector to a concentration of 1 µg/mL in TE buffer.
4. Dilute H5ts125 viral DNA to a concentration of 10 µg/mL in TE buffer.
5. Mix 1 µL of vector and 1 µL of H5ts125 DNA and transform *E. coli* BJ5183-
   competent cells.
6. Screen the ampicillin-resistant colonies using *Xho* I digestion of the mini preparation DNA along with pacAd5 9.2-100 as a positive control and analyze by
   agarose gel electrophoresis.
7. Amplify the positive clone(s) in *E. coli* DH5α cells (pac-ts125 9.2-100).
8. Confirm the presence of the ts125 mutation (no. 22794-tggta-22798 tgAta) by
   DNA sequence analysis.

3.5. Transfection and Generation of Recombinant Viruses

1. One day prior to transfection, plate human 293 cells in 60-mm dishes (1×10^6
   cells/dish) with 5 mL of growth medium containing 10% fetal bovine serum.
2. Digest 20 µg each of pacAd5 9.2-100 and pac-ts125 9.2-100 with *Pac*I, precipitate with ethanol, and dissolve in TE at a concentration of 1 mg/mL.
3. On the day of transfection, mix 4 µg of digested pacAd5 9.2-100 with 4 µg of
   pLendΔE1BQ4 vector, pLendΔE1BQ4-1,2BAX or pLendΔE1BQ4-3,4BAK (to
   generate empty shRNA, BAX-shRNA, or BAK-shRNA expressing virus) with
   carrier salmon-sperm DNA to give a total of 20 µg. Add 125 µL of 2 *M* CaCl$_2$
   and make the volume to 500 µL with sterile water in a 12- × 75-mm snap-cap
   tube. Add 500 µL of 2X HeBS into a 17- × 100-mm snap-cap tube.
4. Transfer slowly the DNA–CaCl$_2$ mixture drop-by-drop to 500 µL 2X HeBS with
   a 1-mL pipet and plastic tip while vortexing HeBS.
5. Add 500 µL of the precipitate to each 60-mm dish with cells and mix gently.
6. Incubate at 37°C for 5 h.

7. Remove the medium with precipitates and feed the cells with fresh medium containing 2% fetal bovine serum.

8. Maintain the transfected cells at 37°C until the cells show cytopathic effect (CPE). Every fourth day feed the cells with fresh medium containing 2% fetal bovine serum. Expect CPE within 10 d of transfection.

9. Repeat the same transfection experiment using pac-ts125 9.2-100 instead of pacAd5 9.2-100 to generate the shRNA-expressing viruses in the temperature-sensitive mutant background. In addition, incubate the transfected cells at 32°C (permissive temperature for ts125) instead of 37°C.

10. Freeze the transfected cells after they show CPE at –80°C.

11. Thaw at room temperature inside the hood and transfer to 10-mL snap-cap tube kept on ice.

12. Sonicate for 3 min and centrifuge for 15 min at 3000*g* in a refrigerated centrifuge.

13. Name the viruses according to their mutation and shRNA expression.
    a. Ad5ΔE1BQ4 (wt background and no shRNA expression).
    b. Ad5ΔE1BQ4-1,2BAX (wt background with BAX shRNA expression at sites 1 and 2).
    c. Ad5ΔE1BQ4-3,4BAK (wt background with BAK shRNA expression at sites 3 and 4).
    d. ts125ΔE1BQ4 (ts125 background and no shRNA expression).
    e. ts125ΔE1BQ4-1,2BAX (ts125 background with BAX shRNA expression at sites 1 and 2).
    f. ts125ΔE1BQ4-3,4BAK (ts125 background with BAK shRNA expression at sites 3 and 4).

14. Store the virus-containing clear supernatant in a –80°C freezer.

15. Prepare large-scale stock and purify by banding in CsCl. Determine the titer by standard plaque assay *(13)*. Titrate the viruses with ts125 mutation in human 293 cells at 32°C.

3.6. Analysis of Target Protein Expression

1. One day prior to infection, plate A549 cells in 60-mm dishes (1×10^6 cells/dish) with 5 mL of growth medium containing 10% fetal bovine serum.

2. Remove medium and add virus (20 PFU/cell) in 0.5 mL of growth medium containing 2% fetal bovine serum.

3. Incubate at 37°C for 1 h with three gentle shakings.

4. Remove the virus-containing medium and add fresh medium containing 2% fetal bovine serum.

5. Incubate the cells infected with the *ts* viruses at 39.5°C and cells infected with non-ts viruses at 37°C.

6. At the end of 12, 24, and 48 h postinfection for the non-ts viruses and 12, 24, 48, 72, and 96 h postinfection for the ts viruses, collect the floating cells in the medium by spinning at 300*g*, resuspend the cell pellet in 50 μL of PBS, and add back to the dish containing the virus-infected cells.

7. Lyse the cells in 450 μL of sample buffer, transfer to a microcentrifuge tube, and store at –20°C until use.
8. Boil the samples for 5 min and load 30 μL of sample onto a 12% polyacrylamide gel and do the Western blot using a standard protocol *(8)* with rabbit polyclonal antibodies specific for human BAX and BAK as primary antibodies and HRP-conjugated goat anti-rabbit IgG as the secondary antibody.
9. Develop the blots using the chemiluminescent detection system.

With the above method, we have successfully constructed various viruses for the suppression of BAX and BAK (**Fig. 2**). The shRNA expression vectors (non-ts125) did not satisfactorily inhibit the expression of BAX and BAK (**Fig. 2, A** and **C**). This could be the result of cell death that occurs within 48 h after infection as well as after inhibition of the shRNA activity by the VA-RNA. A reduction in the levels of BAX and BAK observed at 48 h postinfection might be the result of cell death. In cells infected with the ts shRNA, the expression of BAX and BAK was significantly inhibited after 72 h postinfection (**Fig. 2B** and **D**). The *ts* shRNA vector described may be useful to investigate the role of various cellular genes in Ad-induced apoptosis. The vector may also be useful in studying the role of cellular genes in virus replication. It could be envisioned that vectors expressing shRNA from the E3 region might also be useful for these purposes.

4. Notes

1. The larger plasmids pacAd5 9.2-100 and pac-ts125 9.2-100 should be amplified in *E. coli* DH5α. In HB101 and BJ5183 these plasmids are unstable.
2. The DNA of pacAd5 9.2-100 and pac-ts125 9.2-100 should be digested with *Pac*I restriction enzyme before transfection into 293 cells to facilitate efficient overlap recombination. Transfection of plasmid DNAs without *Pac*I digestion does not result in virus production.
3. After transfection, incubate the pac-ts125 9.2-100 transfected cells at 32°C. Incubation at 37°C does not result in generation of recombinant virus.
4. If pacAd5 9.2-100 and pac-ts125 9.2-100 plasmids are used along with left-end plasmid for generating recombinant virus, generally there is no need to screen the progeny viruses. In our experience, we have not detected any virus that was unexpected. The expected mutation (*ts*125) can be confirmed by PCR using the Hirt supernatant DNA prepared from the infected cells.
5. Virus preparation and titration for the *ts* mutants should be carried out in human 293 cells and at 32°C. For shRNA expression experiment the cells have to be incubated at the nonpermissive temperature of 39.5°C.

Acknowledgments

This work was supported by research grants CA-33616 and CA-84941 from the National Cancer Institute, NIH.

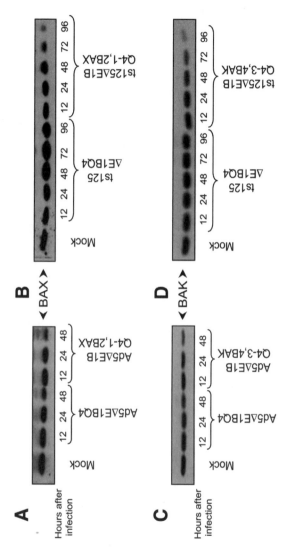

Fig. 2. Suppression of BAX and BAK shRNA expressing Ad vectors. A549 cells were infected (20 PFU/cell) with the shRNA-expressing non-ts vectors (**A** and **C**) and incubated at 37°C. At the indicated times postinfection, cells were lysed in sample buffer and subjected to Western blot analysis using anti-BAX and anti-BAK antibodies. The cells infected with the ts shRNA vectors were incubated at 39.5°C and analyzed for the expression of BAX (**B**) and BAK (**D**) at the indicated time points.

References

1. Dykxhoorn, D. M., Novina, C. D., and Sharp, P. A. (2003) Killing the messenger: short RNAs that silence gene expression. *Nat. Rev. Mol. Cell. Biol.* **4,** 457–467.
2. Meister, G. and Tuschl, T. (2004) Mechanisms of gene silencing by double-stranded RNA. *Nature* **431,** 343–349.
3. Zhao, L. J., Jian, H., and Zhu, H. H. (2003) Specific gene inhibition by adenovirus mediated expression of small interfering RNA. *Gene* **316,** 137–141.
4. Anderson, R. D., Haskell, R. E., Xia, H., Roessler, B. J., and Davidson, B. L. (2000) A simple method for the rapid generation of recombinant adenovirus vectors. *Gene Ther.* **7,** 1034–1038.
5. Anderson, M. G., Haasnoot, P. C. J., Xu, N., Berenjian, S., Berkhout, B., and Akusjarvi, G. (2005) Suppression of RNA interference by adenovirus virus-associated RNA. *J. Virol.* **79,** 9556–9565.
6. Mayer, A. J. and Ginsberg, H. S. (1977) Persistence of type 5 adenovirus DNA in cells transformed by a temperature sensitive mutant, H5ts125. *Proc. Natl. Acad. Sci. USA* **74,** 785–788.
7. Kruijer, W., van Schaik, F. M. A., and Sussenbach, J. S. (1981) Structure and organization of the gene coding for the DNA binding protein of adenovirus type 5. *Nucleic Acids Res.* **9,** 4439–4457.
8. Subramanian, T. and Chinnadurai, G. (2003) Pro-apoptotic activity of transiently expressed BCL-2 occurs independent of BAX and BAK. *J. Cell. Biochem.* **89,** 1102–1114.
9. Sui, G., Soohoo, C., Affar el B., Gay, F., Shi, Y., Forrester, W. C., and Shi, Y. (2002) A DNA vector-based RNAi technology to suppress gene expression in mammalian cells. *Proc. Natl. Acad. Sci. USA* **99,** 5515–5520.
10. Wang, S. and El-Deiry, W. S. (2003) Requirement of p53 targets in chemosensitization of colonic carcinoma to death ligand therapy. *Proc. Natl. Acad. Sci. USA* **100,** 15,095–15,100.
11. Ohtsuka, T., Ryu, H., Minamishima, Y. A., et al. (2004) ASC is a Bax adaptor and regulates the p53-Bax mitochondrial apoptosis pathway. *Nat. Cell Biol.* **6,** 121–128.
12. Chartier, C., Degryse, E., Gantzer, M., Dieterle, A., Pavirani, A., and Mehtali, M. (1996) Efficient generation of recombinant adenovirus vectors by homologous recombination in *Escherichia coli. J. Virol.* **70,** 4805–4810.
13. Tollefson, A. E., Hermiston, T. W., and Wold, W. S. M. (1999) Preparation and titration of CsCl-banded adenovirus stock. In *Adenovirus Methods and Protocols*, (Wold, W. M. S., ed.), Humana, Totowa, NJ, pp. 1–9.

10

Evaluating Apoptosis in Tumor Cells Infected With Adenovirus Expressing p53 Tumor Suppressor Gene

Elizabeth Perron, M. Behzad Zafar, Amanda Pister, Zhen-Guo Wang, Jun Tian, and Prem Seth

Summary

This chapter describes several methods for recognizing apoptosis in tumor cells following infection with a replication-deficient adenovirus expressing the tumor suppressor gene p53. We include cytotoxicity assays and assays of apoptosis, including DNA–nucleosomal DNA fragmentation (DNA laddering), TUNEL, DAPI staining, analysis of the sub-G_1 (subdiploid) population, and degradation of poly(ADP-ribose) polymerase (as assayed by Western blot). Although this is not a comprehensive list of protocols to evaluate apoptosis, we believe that these will cover the majority of conditions of apoptosis that may arise. The chapter also describes the characteristics of each technique, including the advantages and disadvantages of each method.

Key Words: Apoptosis; programmed cell death; adenovirus; p53; nucleosomal DNA degradation; caspases; PARP cleavage; DAPI assay; TUNEL assay; SRB assay; MTS assay.

1. Introduction

In recent years, adenoviral vectors expressing tumor suppressor gene p53 have been proposed for cancer gene therapy *(1–3)*. Earlier work had indicated that tumor cells transduced with either a retrovirus or a plasmid expressing p53 exhibited a reduced growth rate compared to the tumor cells transduced with a control vector *(4)*. However, in these experiments, it was not clear whether this overexpression of p53 was inducing cell cycle arrest and/or apoptosis (programmed cell death). The availability of adenoviral vectors expressing high levels of p53 protein provided an interesting tool to address the question of whether or not the ectopic expression of p53 in tumor cells can induce apoptosis *(5,6)*. In the mid-1990s, our laboratory generated a first-generation aden-

From: *Methods in Molecular Medicine, Vol. 130:*
Adenovirus Methods and Protocols, Second Edition, vol. 1:
Adenoviruses Ad Vectors, Quantitation, and Animal Models
Edited by: W. S. M. Wold and A. E. Tollefson © Humana Press Inc., Totowa, NJ

oviral vector expressing human p53 tumor suppressor gene (AdWTp53) *(5)*. We demonstrated that infection of tumor cells with AdWTp53 could induce apoptosis in these cells *(5)*. In this chapter, various methods employed to evaluate AdWTp53-mediated apoptosis are described *(5,7–13)*. These methods include (1) evaluation of cytotoxic effects expected to be observed in cells undergoing apoptosis, (2) nucleosomal degradation resulting in DNA laddering, (3) increased end-labeling of DNA (TUNEL assays), (4) increased binding of DNA with dyes such as DAPI, (5) appearance of sub-G_1 peak (subdiploid DNA) in flow cytometric analysis, and (6) caspase activity resulting in degradation of several cellular proteins including poly(ADP-ribose) polymerase (PARP). Although we have used these methods for exploration of AdWTp53-induced apoptosis in tumor cells, the techniques can be used for any number of other purposes needing apoptosis testing.

2. Materials

Listed below are the necessary supplies for the protocols described later. The materials are organized in the order in which they are required in accordance with the given protocols.

1. Stock of tumor cells (number, as wells as tissue source, may vary).
2. 96-Well plates (Corning, Inc., Corning, NY).
3. Incubator held at 37°C.
4. Location for samples to be held at 4°C for extended periods of time.
5. Iscove's minimal essential medium (IMEM) containing 10% fetal bovine serum (FBS) (Sigma, St. Louis, MO).
6. Adenoviral vectors (AdWTp53 and AdControl):
   a. AdWT53: replication-deficient adenovirus expressing *p53* gene.
   b. AdControl: replication-deficient adenovirus devoid of transgene.
7. Trichloroacetic acid (Sigma).
8. Sulforhodamine B (SRB): 0.4% (w/v) (Sigma).
9. BioKinetics Reader EL340 (Bio-Tek, Inc., Burlingame, CA).
10. 3-(4,5-Dimethylthiazol-2-yl)-5-(3-carboxymethoxyphenyl)-2-(4-sulfophenyl)-2H tetrazolium (MTS) reagent (Promega, Madison, WI).
11. 10-cm Dishes (Corning, Inc.).
12. Centrifuge (RT-6000B, Du Pont, Boston, MA).
13. Cold phosphate-buffered saline (PBS) (Sigma).
14. 10 m*M* Trichloroacetic acid (ICN Biomedicals, Inc., Aurora, OH).
15. 10 m*M* Ethylene diamine tetraacetic acid (EDTA) disodium, pH 7.4 (Research Genetics, Huntsville, AL).
16. 0.2 mg/mL Proteinase K (Boehringer Mannheim, Mannheim, Germany).
17. Water bath at 55°C.
18. Sodium chloride (Sigma).
19. Isopropanol (Sigma).

20. Beckman microfuge (Beckman Coulter, Inc., Fullerton, CA).
21. DNase-free RNase (Ambion, Inc., Austin, TX).
22. Agarose gel (2%) (ISC BioExpress, Kaysville, UT).
23. Trypsin (Invitrogen, Carlsbad, CA).
24. Bovine serum albumin (BSA) (Pierce, Rockford, IL).
25. 4, 6-Diamino-2-phenylindole (DAPI) (store in the dark at 4°C) (Sigma).
26. 0.05 M Phosphate buffer, pH 7.2 (Sigma).
27. MESTAIN apoptosis kit (Medical & Biological Laboratories, Co., Ltd., Nagoya, Japan).
28. Propidium iodide solution: 50 µg/mL in 0.1% sodium citrate.
29. FACSCalibur instrument (Becton Dickinson, Franklin Lakes, NJ).
30. Fluorescent microscope (Eclipse TE200, Nikon, Melville, NY).
31. 6-cm Dishes (Corning, Inc.).
32. 1X Sodium dodecyl sulfate–polyacrylamide gel electrophoresis (SDS-PAGE) buffer: 62 mM Tris-HCl, pH 6.8, 2 mM EDTA, 15% sucrose, 10% glycerol, 3% SDS, and 0.7 M 2-mercaptoethanol(Sigma).
33. SDS-PAGE gels (Fisher Scientific, Fairlawn, NJ).
34. Nitrocellulose filters (Nikon).
35. Tris-buffered saline, containing 5% dried milk and 0.1% Tween-20 (Sigma).
36. PARP antibody (Oncogene Science, Uniondale, NY).
37. Horseradish peroxidase (Sigma).

3. Methods

Apoptosis is a form of cell death in which a programmed sequence of events leads to the maintenance of the tissue by eliminating old, unnecessary, and unhealthy cells *(14,15)* (*see* **Note 1**). There are several procedures that detect programmed cell death. Because all of the protocols have a similar purpose, it is important to recognize the advantages and disadvantages of each method (found under the note section for each of the methods). Below are several different methods for recognizing apoptosis in AdWTp53-infected cells. This is not an exhaustive list, but we believe that these methods will cover a majority of conditions that may arise. We have included the SRB assay and MTS assay (**Subheading 3.1.**); DNA–nucleosomal DNA fragmentation (DNA laddering) assay (**Subheading 3.2.**); TUNEL assay (**Subheading 3.3.**); DAPI staining assay (**Subheading 3.4.**); analysis of the sub-G_1 (subdiploid) population (**Subheading 3.5.**); and protein degradation of PARP cleavage (by Western blot) (**Subheading 3.6.**).

3.1. Cytotoxicity Assays

Because AdWTp53-mediated cytotoxicity is a good indicator of apoptosis, the following two methods can be utilized to measure this cytotoxic effect in AdWTp53-infected tumor cells.

3.1.1. SRB Toxicity Assay

The SRB assay can be used to measure AdWT53-induced cytotoxicity in tumor cells. This assay is based on the ability of the SRB dye to bind to proteins under mild acidic conditions. The bound complexes can be extracted from cells and the soluble red color measured using a spectrophotometer *(5,16–18)* (*see* **Note 2**).

1. Plate 500 cells per well in 96-well plates.
2. Incubate overnight at 37°C in IMEM containing 10% FBS.
3. The following day, infect the cells with increasing dosages $(0-10^4$ plaque-forming units [PFU] per cell) of adenoviral vectors (AdWTp53 or AdControl).
4. Incubate for 7 days at 37°C.
5. After incubation, fix the cells with 10% trichloroacetic acid (final concentration) and stain with 0.4% SRB.
6. Measure the absorbance at 564 nm by using a BioKinetics Reader EL340. The IC_{50} dose is the amount of adenoviral vector required to inhibit cell growth by 50%.

3.1.2. MTS Assay

The MTS assay is another type of toxicity assay. It determines AdWT53-induced cytotoxicity as a measure of cellular metabolic activity *(10)*. Experiments can be done in order to measure metabolic function through mitochondrial activity (through MTS metabolism) (*see* **Note 2**).

1. Plate 500 cells per well in 96-well plates; incubate the cells for 24 h at 37°C.
2. Infect the cells with the adenovirus (AdWTp53 or AdControl) $(0-10^4$ PFU per cell) by adding the adenovirus to the wells.
3. Incubate for an additional 48 h at 37°C.
4. Add the MTS reagent according to the manufacturer's instructions and read the O.D. on the Bio Kinetics Reader.
5. Calculate the toxicity percentage from the surviving cells; assume 100% survival for uninfected cells.

3.2. DNA–Nucleosomal DNA Fragmentation (DNA Laddering) Assay

A major characteristic of the apoptotic process is DNA fragmentation caused by endonucleases activated during the apoptotic process. Because cleavage of DNA into discrete fragments characteristically occurs prior to absolute membrane disintegration, a procedure that measures DNA fragmentation is well suited for the determination of cell death *(5,9,19,20)*. (DNA fragmentation quantification will be described in further detail in **Subheading 3.3.**)

DNA fragmentation can be observed in AdWTp53-infected cells through analysis by gel electrophoresis; the end product is a typical DNA ladder pattern. Ordinarily, DNA is located exclusively in the nucleus in high molecular-weight form. Following degradation of the DNA by endonucleases, apoptotic cells have lower than normal concentrations of high molecular-weight DNA in the nucleus and some level of low molecular-weight DNA present in the nucleus (*see* **Note 3**). Low molecular-weight DNA is also found in the cytoplasm as a result of the leakiness of the apoptotic cell.

1. Plate 2×10^6 cells in 10-cm dishes.
2. On the next day, infect cells with adenoviral vectors (AdWTp53 or AdControl) (100 PFU/cell) and incubate at 37°C for 48 h.
3. Collect adherent and floating cells and centrifuge together at 1800g for 5 min; aspirate the supernatant.
4. Rinse the cell pellets with cold PBS.
5. Lyse pellets in 1 mL of 10 mM Tris-HCl, pH 7.4, 10 mM EDTA, and 0.2 mg/mL proteinase K.
6. Incubate the samples at 55°C in a water bath for 5 h.
7. Add sodium chloride (to a final concentration of 1 M).
8. Incubate at 4°C for 14 h.
9. Add an equal volume of isopropanol.
10. Centrifuge at 15,000g for 30 min using the Beckman microfuge.
11. Collect the supernatant.
12. Add DNase-free RNase (final concentration of 100 mg/mL).
13. Incubate at 37°C for 1 h.
14. Evaluate the DNA by electrophoresis on an agarose gel (2%).

3.3. TUNEL Assay

This procedure explores AdWTp53-mediated DNA fragmentation in single cells and is also known as *in situ* cell death enzyme-linked immunosorbent assay (ELISA). The procedure allows for further examination of nuclear changes, as it detects strand breaks within the DNA. The enzyme Tdt (terminal-deoxy-transferase) is used to add dUTPs to the broken ends of the fragmented DNA; these can then be detected through the use of antibodies with fluorescent labels *(8,12,13,21)*.

TUNEL has several applications, which include (1) detection of single apoptotic cells in frozen and formalin-fixed tissue sections for basic research and routine pathology, (2) determination of the sensitivity of malignant cells to drug-induced apoptosis in cancer research and clinical oncology, and (3) typing of cells that are undergoing cell death in heterogeneous populations by double-staining procedures (*see* **Note 4**).

1. Plate 2×10^4 cells in 35-mm dishes; incubate the cells for 24 h at 37°C.
2. Infect cells with AdWT53 or AdControl (100 PFU/cell) for 48 h.
3. Collect samples by trypsin-EDTA, and wash twice with PBS.
4. Perform TUNEL using the MESTAIN apoptosis kit according to the manufacturer's instructions.
5. Perform FACS analysis using the FACSCalibur instrument.

3.4. DAPI Staining Assay

DAPI staining is often one of the first assays performed for detection of apoptotic cells. This procedure is also described as a dye exclusion method. This means that intact and damaged plasma membranes can be discriminated by staining. DAPI is known primarily for its ability to form fluorescent complexes with natural double-bonded strands of DNA, showing its fluorescence at those areas that contain hydrogen bonds *(8,9)*.

Cells with injured plasma membranes are permeable to the stain, whereas healthy cells, whose plasma membrane is not permeable, will not show staining. In AdWTp53-infected tumor cells, when DAPI binds to DNA its fluorescence is strongly enhanced. The stained and unstained cells are then counted. The apoptotic cells can be further distinguished from normal cells by membrane blebbing and formation of apoptotic bodies, in addition to DAPI staining (*see* **Note 5**).

1. Plate 2×10^4 cells in 35-mm dishes; incubate the cells for 24 h at 37°C.
2. Incubate cells with adenoviruses (AdWTp53 or AdControl, 100 PFU/cell) for 48 h, then scrape the cells off the plates.
3. Wash the cell cultures with PBS and harvest by trypsinization.
4. Resuspend 5×10^5 cells in 50 mL of PBS.
5. Add 10 mL of 22% BSA to sample and pipet cell/BSA suspension into the bottom of a cytofunnel mounted with a microscope slide into the cytocentrifuge and spin at 5000g for 5 min.
6. Dry slides for 30 min at room temperature (to allow evaporation of much of excess liquid) and wash twice with PBS.
7. Add 200 µL of DAPI (2.5 µg/mL in 0.05 M phosphate buffer, pH 7.2) and incubate for 30 min at room temperature.
8. Wash the cells twice with PBS.
9. Photograph the cells using a fluorescent microscope.
10. Store samples stained with DAPI in the dark at 4°C.

3.5. Analysis of the Sub-G_1 (Subdiploid) Population

This analysis allows for the quantification, upon staining, of the percentage of the subdiploid population. Cells with lower DNA staining than that of G_1-cells (sub-G_1 peaks) are considered apoptotic *(8,22)*. The sub-G_1 analysis is performed following the rinse and staining processes (*see* **Note 6**).

1. Plate 5×10^4 tumor cells in a 35-mm dish.
2. On the next day, infect the cells with AdWTp53 (multiplicity of infection [MOI] of 100 PFU/cell for 48 h).
3. Add 1 mL of trypsin and prepare single-cell suspensions.
4. Wash the cells and resuspend in 0.5 mL of PBS.
5. Fix the cells, using 2% paraformaldehyde as fixative (*see* **Note 7**).

6. Wash the cells and resuspend in PBS to a density of $1-2 \times 10^6$/mL.
7. Add propidium iodide in PBS to a final concentration of 50 µg/mL.
8. Stain samples for 1–24 h (depending on cell type).
9. Measure the fluorescent activity of propidium iodide-stained DNA of permeabilized and fixed cells on the FACSCalibur instrument.
10. Use the following equation to evaluate the sub-G_1 population of cells:

$$\% \text{ Sub-}G_1 \text{ population} = [(\text{no. of cells below } G_1) / (\text{total no. analyzed})] \times 100$$

3.6. Protein Degradation of PARP Cleavage by Western Blot

During apoptosis, proteins within the cell are cleaved and denatured by caspases (acronym for cys-containing asp-specific proteases). Once caspases are activated, the apoptotic process is initiated, including (1) condensing of the cell nucleus and formation of micronuclei, (2) formation of membrane-bound apoptotic bodies, and (3) DNA fragmentation (fragments containing multiple nucleosomes). Evidence of a link between caspases and apoptosis activation stems from the ability of specific caspase inhibitors to block programmed cell death.

Caspase-3 is the most common caspase and is shown to target and cleave a sequence that is specific to PARP (an enzyme involved in DNA repair and genomic maintenance) *(23–25)*. This assay allows examination of PARP cleavage by using an anti-PARP antibody *(8,10,26)*. The crude cell extract is used for Western blot analysis. Cleaved PARP fragments are detected with the antibody through either an achromogenic or a chemiluminescent reaction (*see* **Note 8**).

1. Plate 5×10^5 cells in 6-cm dishes and incubate at 37°C.
2. The next day, infect the cells with adenovirus (AdWTp53 or AdControl) for 24 h.
3. Wash cells three times with ice-cold PBS.
4. Collect the cells by gently scraping the bottom of the dishes and resuspend the cells in 1 mL 1X SDS-PAGE buffer. A Bradford Assay is performed for protein estimation using GelCode® Blue Stain Reagent (Pierce; this reagent is designed to work with SDS-PAGE buffer).
5. Boil the solution for 10 min.
6. Perform electrophoresis on equal amounts (15 to 50 µ g) of denatured protein on SDS-polyacrylamide gels and transfer to nitrocellulose membrane filters.
7. Block membrane filters with Tris-buffered saline (TBS) containing 5% dried milk and 0.1% Tween-20.
8. Probe the blots with PARP antibody.
9. Following incubation with the primary antibodies, wash the blots with TBS containing 0.1% Tween-20.
10. Incubate the blots with horseradish peroxidase-conjugated secondary antibody.
11. Following incubation, wash the blots with TBS/0.1% Tween-20.
12. Detect the specific complexes by the chemiluminescent technique.

4. Notes

1. Apoptosis is the most frequent form of programmed cell death, but other nonapoptotic types of cell death do occur. Apoptosis is distinguished from necrosis, which is associated with the external damage to cells; apoptotic cells actively participate in their own destruction. Cells that are damaged by injury, such as mechanical damage or exposure to toxic chemicals, undergo a series of changes characteristic of necrosis. These cells will swell, and their cellular contents, including organelles and cytoplasm, will leak out, leading to inflammation of any surrounding tissues.

2. SRB and MTS assays are inexpensive assays that can be used for screening purposes, but they should not be used exclusively. Thus, their purpose should be mainly as a screening assay. Because these assays can be done in 96-well plates, many assays can be completed at one time, making them excellent screening methods.

3. The activation of endonucleases, which cleave the cellular DNA, occurs at a late stage of apoptosis and leads to the laddering fragments seen in this technique. The presence of DNA in a typical ladder pattern, with approx 200 base pair (bp) repeats, is a good marker for apoptosis. Although DNA fragmentation analysis is simple, convenient, and generally able to produce good results, it is very laborious to process large numbers of samples and it provides no quantitative analysis.

4. Endonuclease activation, which is examined in this technique, results in cleavage of the DNA of the cell during one of the later stages of apoptosis. The protocol describes plating the cells on a slide, but they may also be plated in a 6-cm dish. This procedure can be technically demanding for some and does have the disadvantage of measuring an event that occurs intermediate or late in the apoptotic process. However, its advantages far exceed its disadvantages. TUNEL has become a very popular assay because it is a precise, fast, nonradioactive technique to detect and quantify apoptotic cell death at the single-cell level in all cell types.

5. DAPI staining, as described here, examines changes that have occurred to the cellular membrane. Membrane alterations are an intermediate step in the apoptotic process, so not all cells undergoing apoptosis may be stained. Also, difficulty in phenotyping the cells and application to only suspended cells can be seen as disadvantages with this procedure. However, DAPI staining has been found to be an easy assay to perform. Other stains besides DAPI may be used, including nonfluorescent dyes. When using dyes other than DAPI, a similar protocol may be followed, but some alterations may need to be made. DAPI and DAPI-stained samples must not be exposed to light; store them in the dark or wrap in foil.

6. An issue with this technique is that apoptotic G_2-phase cells also exhibit reduced DNA content, which could resemble the DNA content of a G_1-cell. You are looking for a specific G_1 peak, not simply any data showing this peak. Thus, some apoptotic cells may not be classified as apoptotic, and this could result in an underestimation of the apoptotic cellular population. Also, this analysis cannot be performed on apoptotic cells that have not undergone DNA fragmentation. Moreover, no information at the single-cell level can be obtained with this technique.

7. For this step, use fixative chilled to –20°C. Add fixative dropwise to the cell suspension until the appropriate concentration is achieved, with constant shaking. One hour in the fixative is sufficient for most cells. Longer times are usually not harmful. Keep samples in the cold during PI staining and running of samples (flow cytometry).

8. Because caspase-3 is activated during the earlier stages of apoptosis, this method is an early marker of apoptosis. The blocking step may be carried out overnight at 4°C, if desired. If the substrate solution develops precipitate during storage at 4°C, warm it to room temperature and mix; a sonicating water bath may also be used. A small amount of precipitate in these solutions will not harm the performance of the product. Keep substrate solution away from open flames and avoid contact with skin, eyes, and mouth. The Annexin V assay is an alternative early-detection technique, not described here, that follows the early apoptotic changes of the plasma membrane. The Annexin V assay can be performed, but we feel that it is one of the less accurate assays. Many substrates other than PARP can be used; when looking for another substrate, a similar protocol may be followed.

References

1. Seth, P. (2005) Vector-mediated cancer gene therapy: an overview. *Cancer Biol. Ther.* **4,** 512–517.
2. Seth, P. (2000) Pre-clinical studies with tumor suppressor genes. *Adv. Exp. Med. Biol.* **465,** 183–192.
3. Roth, J. A. and Grammer, S. F. (2003) Tumor suppressor gene therapy. *Method Mol. Biol.* **223,** 577–598.
4. Casey, G., Lo-Hsueh, M., Lopez, M. E., Vogelstein, B., and Stanbridge, E. J. (1991) Growth suppression of human breast cancer cells by the introduction of a wild-type p53 gene. *Oncogene* **6,** 1791–1797.
5. Katayose, D., Gudas, J., Nguyen, H., Srivastava, S., Cowan, K. H., and Seth, P. (1995) Cytotoxic effects of adenovirus-mediated wild-type p53 protein expression in normal and tumor mammary epithelial cells. *Clin. Cancer. Res.* **1,** 889–897.
6. Zhang, W. W., Alemany, R., Wang, J., Koch, P. E., Ordonez, N. G., and Roth, J. A. (1995) Safety evaluation of Ad5CMV-p53 in vitro and in vivo. *Hum. Gene Ther.* **6,** 155–164.
7. Katayose, D., Wersto, R., Cowan, K. H., and Seth, P. (1995) Effects of a recombinant adenovirus expressing WAF1/Cip1 on cell growth, cell cycle, and apoptosis. *Cell Growth Differ.* **6,** 1207–1212.
8. Katayose, Y., Kim, M., Rakkar, A. N., Li, Z., Cowan, K. H., and Seth, P. (1997) Promoting apoptosis: a novel activity associated with the cyclin-dependent kinase inhibitor p27. *Cancer Res.* **57,** 5441–5445.
9. Li, Z., Rakkar, A., Katayose, Y., et al. (1998) Efficacy of multiple administrations of a recombinant adenovirus expressing wild-type p53 in an immune-competent mouse tumor model. *Gene Ther.* **5,** 605–613.
10. Rakkar, A. N., Katayose, Y., Kim, M., et al. (1999) A novel adenoviral vector expressing human Fas/CD95/APO-1 enhances p53-mediated apoptosis. *Cell Death Differ.* **6,** 326–333.

11. Slack, R. S., Belliveau, D. J., Rosenberg, M., et al. (1996) Adenovirus-mediated gene transfer of the tumor suppressor, p53, induces apoptosis in postmitotic neurons. *J. Cell Biol.* **135**, 1085–1096.

12. Turturro, F., Frist, A. Y., Arnold, M. D., Pal, A., Cook, G. A., and Seth, P. (2001) Comparison of the effects of recombinant adenovirus-mediated expression of wild-type p53 and p27Kip1 on cell cycle and apoptosis in SUDHL-1 cells derived from anaplastic large cell lymphoma. *Leukemia* **15**, 1225–1231.

13. Rakkar, A. N., Li, Z., Katayose, Y., Kim, M., Cowan, K. H., and Seth, P. (1998) Adenoviral expression of the cyclin-dependent kinase inhibitor p27Kip1: a strategy for breast cancer gene therapy. *J. Natl. Cancer Inst.* **90**, 1836–1838.

14. Brown, J. M. and Attardi, L. D. (2005) The role of apoptosis in cancer development and treatment response. *Nat. Rev. Cancer* **5**, 231–237.

15. Jin, Z. and El-Deiry, W. S. (2005) Overview of cell death signaling pathways. *Cancer Biol. Ther.* **4**, 139–163.

16. Garbin, F., Eckert, K., and Maurer, H. R. (1994) Evaluation of the MTT and SRB assays for testing LAK cell-mediated growth inhibition of various adherent and non-adherent tumor target cells. *J. Immunol. Methods* **170**, 269–271.

17. Perez, R. P., Godwin, A. K., Handel, L. M., and Hamilton, T. C. (1993) A comparison of clonogenic, microtetrazolium and sulforhodamine B assays for determination of cisplatin cytotoxicity in human ovarian carcinoma cell lines. *Eur. J. Cancer* **29A**, 395–399.

18. Voigt, W. (2005) Sulforhodamine B assay and chemosensitivity. *Methods Mol. Med.* **110**, 39–48.

19. Ajiro, K. (2000) Histone H2B phosphorylation in mammalian apoptotic cells. An association with DNA fragmentation. *J. Biol. Chem.* **275**, 439–443.

20. Matassov, D., Kagan, T., Leblanc, J., Sikorska, M., and Akeri, Z. (2004) Measurement of apoptosis by DNA fragmentation. *Methods Mol. Biol.* **282**, 1–17.

21. Maciorowski, Z., Klijanienko, J., Padoy, E., et al. (2001) Comparative image and flow cytometric TUNEL analysis of fine needle samples of breast carcinoma. *Cytometry* **46**, 150–156.

22. Charlot, J. F., Pretet, J. L., Haughey, C., and Mougin, C. (2004) Mitochondrial translocation of p53 and mitochondrial membrane potential (Delta Psi m) dissipation are early events in staurosporine-induced apoptosis of wild type and mutated p53 epithelial cells. *Apoptosis* **9**, 333–343.

23. Yang, X., Stennicke, H. R., Wang, B., et al. (1998) Granzyme B mimics apical caspases. Description of a unified pathway for trans-activation of executioner caspase-3 and -7. *J. Biol. Chem.* **273**, 34,278–34,283.

24. Pinkoski, M. J., Hobman, M., Heibein, J. A., et al. (1998) Entry and trafficking of granzyme B in target cells during granzyme B-perforin-mediated apoptosis. *Blood* **92**, 1044–1054.

25. Wolf, B. B. and Green, D. R. (1999) Suicidal tendencies: apoptotic cell death by caspase family proteinases. *J. Biol. Chem.* **274**, 20,049–20,052.

26. Bharti, A. C., Takada, Y., and Aggarwal, B. B. (2005) PARP cleavage and caspase activity to assess chemosensitivity. *Methods Mol. Med.* **111**, 69–78.

11

Growth and Purification of Enteric Adenovirus Type 40

Vivien Mautner

Summary

The enteric adenoviruses of subgroup F (Ad40 and Ad41) pose some special problems of cultivation, as they cannot be readily passaged in many of the cell types used to propagate the more commonly used subgroup C serotypes (Ad2 and Ad5) and there is no standard plaque assay. Methods to propagate Ad40 in complementing cell lines and to evaluate infectivity and particle number are presented in this chapter.

Key Words: Enteric adenovirus type 40; fluorescent focus assay; Hirt extraction; virus particle count; PicoGreen DNA quantitation.

1. Introduction

The enteric adenoviruses of subgroup F (Ad40 and Ad41) pose some special problems of cultivation, as they cannot be readily passaged in many of the cell types used to propagate the more commonly used subgroup C serotypes (Ad2 and Ad5), and there is no standard plaque assay *(1,2)*. For the purposes of this chapter, I will describe the methods developed in my laboratory to passage Ad40 in 293 and KB16 cells, to assay the virus by a variety of methods, to purify the virus by cesium chloride density gradient centrifugation, and to obtain viral DNA. Most of the methods have been derived from those developed for Ad5, but with constraints imposed by the more fastidious nature of the enteric adenoviruses.

Ad40 strain Dugan *(3)* is available from the American Type Culture Collection (ATCC, Rockville, MD). In my laboratory, the virus was passaged nine times in KB16 cells *(4)*, using a 1:10 dilution of the total yield at each step to provide a p9 stock with a titer by fluorescent focus assay of 6×10^6 FFU/mL and a particle count by electron microscopy of 2×10^{10} particles/mL. This virus was used for all experimental work, and a virus stock grown from this source

From: *Methods in Molecular Medicine, Vol. 130:*
Adenovirus Methods and Protocols, Second Edition, vol. 1:
Adenoviruses Ad Vectors, Quantitation, and Animal Models
Edited by: W. S. M. Wold and A. E. Tollefson © Humana Press Inc., Totowa, NJ

was used to make DNA for cloning and sequencing of the complete Ad40 genome (5) (Genbank accession number L19443). The virion DNA was shotgun cloned into bacteriophage M13mp19 and sequence data generated randomly; from the proportion of nonviral sequences cloned and sequenced, the virion-derived DNA was estimated to be 98.6% pure.

In order to generate substantial amounts of Ad40, we attempted to passage virus seed stocks at a very low multiplicity of infection, as is the procedure for Ad5, but it became apparent that seed stocks diluted beyond 1:30 failed to propagate. We therefore routinely maintain virus stocks by serial passage at 1:10 dilutions and check the yield of virus at each step by comparing levels of virion packaged DNA. It is also important to check the restriction profile for inadvertent contamination with other serotypes; if other adenoviruses are used in the laboratory, this is not a trivial consideration. A comprehensive survey of the restriction profiles of the human adenovirus serotypes (Ad1 to Ad41) is available (6).

Although it has been reported that A549 cells will support plaque formation of the Sapporo strain of Ad40 (7), the Dugan strain cannot be plaqued in 293 cells, so a fluorescent focus assay has been used to estimate the number of infected cells and thence the titer of virus in a stock. For Ad5, the focus forming unit (FFU)-to-plaque forming unit (PFU) ratio is approx 10:1, and the particle-to-PFU ratio ranges from 10 to 100. For Ad40 passaged in 293 or KB16 cells, we find a particle-to-infectivity ratio in the order of 10^3; and although the reason for this large difference is not established, a number of explanations have been suggested (1,8,9).

2. Materials

1. Ad40 strain Dugan: ATCC VR-931 (2).
2. 293 cells: ATCC CRL-1573 (human embryonic primary kidney cells transformed with sheared Ad5 DNA, expressing E1a+b) (10).
3. KB16 cells: (KB epidermoid-carcinoma continuous cell line (ATCC CCL-17) further transformed with Ad2 E1a+b DNA) (4).
4. Dulbecco's modified Eagle's medium (DMEM): Gibco-BRL, Gaithersburg, MD, cat. no. 041-01965 M (without sodium pyruvate, with 4500 mg/L glucose). Store at 4°C.
   a. Pen/strep solution: Gibco-BRL, cat. no. 043-05140 H (1000 IU/mL penicillin, 1000 µg/mL streptomycin). Aliquot and store at –20°C.
   b. L-Glutamine: Gibco-BRL, cat. no. 043-05030 H (200 mM). Aliquot and store at –20°C.
   c. Fetal calf serum (FCS): heat-treat at 56°C for 30–45 min. Aliquot and store at –20°C.
5. Cell-propagation medium: DMEM (500 mL), pen/strep solution (5 mL), L-glutamine (10 mL), FCS (50 mL).

6. Infection medium: DMEM (500 mL), pen/strep solution (5 mL), L-glutamine (10 mL), FCS (2.5 mL). Working stock should be stored at 4°C and used within 10 d.

7. Dulbecco's phosphate-buffered saline (PBS) (complete): 137 mM NaCl, 2.7 mM KCl, 8.0 mM NaH$_2$PO$_4$, 2.4 mM KH$_2$PO$_4$, 6.8 mM CaCl$_2$, 4.9 mM MgCl$_2$.
   For complete PBS use 8 parts solution A, 1 part solution B, and 1 part solution C.
   a. Solution A: NaCl (10 g), KCl (0.25 g), Na$_2$HPO$_4$ (1.43 g), KH$_2$PO$_4$ (0.41 g), pH 7.2, per 1 L H$_2$O.
   b. Solution B: CaCl$_2$·2H$_2$O (1 g) per 1 L.
   c. Solution C: MgCl$_2$·6H$_2$O (1 g) per 1 L.
   Store sterile solutions at 4°C. Make up fresh as required.

8. PBSa: Solution A *(see* **item 7a**).

9. Tris-buffered saline (TBS): 140 mM NaCl, 30 mM KCl, 28 mM Na$_2$HPO$_4$, 25 mM Tris-HCl, glucose (1 mg/mL), pH 7.0. Store sterile at 4°C.

10. Trypsin/versene (T/V): 1 vol trypsin:4 vol versene.
    a. Trypsin: 0.25% in TBS (store sterile at –20°C).
    b. Versene: 0.6 mM ethylene diamine tetraacetic acid (EDTA) in PBSa, 0.002% phenol red (store sterile at 4°C).

11. Tris/EDTA (T/E): 10 mM Tris-HCl, pH 8.0, 1 mM EDTA. Store sterile at room temperature.

12. CsCl-gradient solutions: For a density of 1.45, dissolve 20.5 g CsCl in 2.9 mL 0.5 M Tris-HCl, pH 7.2, and 25.8 mL H$_2$O. For a density of 1.32, dissolve 32.0 g CsCl in 6.8 mL 0.5 M Tris-HCl, pH 7.9, and 61.2 mL H$_2$O. Filter through Whatman no. 1 paper and store at room temperature. The density of the CsCl solutions can be checked on a refractometer.

13. 40% Glycerol: glycerol (40 g), 0.5 M Tris-HCl, pH 7.9 (2.0 mL), 0.2 M EDTA (0.5 mL), H$_2$O to 100 mL. Use sterile reagents, store at room temperature.

14. 80% Glycerol: glycerol (80 g), 0.5 M Tris-HCl, pH 7.9 (2.0 mL), 0.2 M EDTA (0.5 mL), H$_2$O to 100 mL. Use sterile reagents, store at room temperature.

15. Hirt buffer: 10 mM Tris-HCl, pH 7.9, 10 mM EDTA, 0.6% (w/v) SDS. Filter-sterilize and store at room temperature.

16. Proteinase K: Type XXVIII from *Tritrachium album* (Sigma, St. Louis, MO, cat. no. P-4914); stock solution 10 mg/mL in H$_2$O. Store in aliquots at 4°C.

17. Gelvatol: gelvatol 20-30 from Cairns Chemicals, Bucks, UK.
    a. Dissolve 20 g gelvatol in 80 mL: 140 mM NaCl, 10 mM KH$_2$PO$_4$, 10 mM Na$_2$HPO$_4$·12H$_2$O. Shake to dissolve overnight at room temperature.
    b. Add 40 mL glycerol; shake to dissolve overnight at room temperature.
    c. Centrifuge 15,000g 10 min at room temperature. Collect supernatant and check that pH is 6.0–7.0.
    d. Store aliquots at 4°C.

18. Citifluor (Citifluor, London).

19. 90% Methanol in H$_2$O.

20. Antibodies:

a. Group-specific adenovirus antibody, e.g., murine monoclonal antibody such as 10/5.1.2 NCL-Adeno from Novocastra, Newcastle, UK.
b. Fluorescein-conjugated goat anti-mouse IgG.
c. Rhodamine-conjugated bovine serum albumin (BSA) (Nordic Immunological Laboratories, PO Box 22 5000 AA Tilburg, The Netherlands).

21. Latex beads: 225-nm diameter beads, 1.6×10^{11}/mL (Agar Scientific, Stanstead, Essex, UK, cat. no. S 130-4).
22. PicoGreen dsDNA Quantitation Kit™ (Invitrogen/Molecular Probes cat. no. P-7589 and P-11496).
23. Bio-Rad Protein Assay dye concentrate (Hercules, CA, cat. no. 500-0006).
24. Stain for electron microscope: sodium silicotungstate, pH 7.0, or phosphotungstic acid, pH 7.0 (Agar Scientific).

3. Methods

3.1. Propagating Cells

Cell stocks are maintained in 80-cm^2 tissue-culture flasks and passaged on a regular basis before they become confluent; local conditions and the split used will determine the frequency. In my hands, 293 cells can be split 1:10 every 7 d to maintain the stock. One 80% confluent flask yields approx 2×10^7 cells, which will provide 20 60-mm plates seeded at 10^6 cells/plate. These should reach 80% confluence in 2 d. KB16 cells are split 1:4 every 3–4 d; they are less tolerant of a higher dilution.

1. Remove medium from the cells, add 5 mL prewarmed T/V, gently wash cells until pH becomes acid. Discard T/V, add a further 5 mL and incubate at 37°C until the cells detach (usually a maximum of 5 min; do not leave longer than necessary).
2. Bang the flask to detach all the cells; allow to drain to base of flask.
3. Add 20 mL prewarmed medium to the flask (serum in the medium neutralizes the trypsin).
4. Disperse the cells by pipetting up and down three times with a 10-mL pipet.
5. Seed flasks as required; e.g., 2.5 mL cell stock + 20 mL medium for a 1:10 split.
6. Regularly test all cell lines for mycoplasma, e.g., once every 2 mo. Passage newly recovered cells four times in antibiotic-free medium before testing.

3.2. Preparing Plates for Infection

1. Trypsinize cells from a 90% confluent 80-cm^2 flask as described above and resuspend in 25–30 mL of medium.
2. Count an aliquot in a hemacytometer, and plate $1–2 \times 10^6$ cells per 60-mm plate in 4–5 mL medium. The exact number of cells will be determined by the growth characteristics of the cells, and whether they are required at 1 or 2 d postplating.

3.3. Preparation of Virus Seed Stock (see Note 1)

Ad40 seed stock should be generated in 60-mm plates by serial passage of 1:10 dilutions. This is preferable to using large flasks, as spot contamination of individual plates can be readily recognized and stocks discarded accordingly.

1. Remove medium from subconfluent monolayers of 293 or KB16 cells in 60-mm plates.
2. Infect with virus diluted in TBS to give approx 1 FFU/cell in 0.1 mL/plate (generally a 1:10 dilution).
3. Adsorb at 37°C for 60–90 min. Rock plates gently at 20-min intervals.
4. Add 5 mL medium + 0.5% FCS.
5. Cells should take 5–7 d to show a good cytopathic effect (CPE) (Ad40 has less tendency than Ad5 to produce a "bunch of grapes" effect, but individual cells do round up).
6. When a good CPE is apparent, scrape cells into the medium with a pastette or teflon scraper and pipet cells gently to disperse clumps.
7. Pool cell suspensions and centrifuge at 900g for 10 min at 4°C.
8. Drain pellets and resuspend in approx 0.25 mL TBS/60-mm plate.
9. To release virus, freeze–thaw three times, i.e., freeze in dry ice or liquid nitrogen, and thaw at 37°C in water bath.
10. Centrifuge at 900g for 10 min at 4°C. Small volumes can be centrifuged in Eppendorf tubes at 1500g for 1 min.
11. Aliquot supernatant; store at –20°C.

3.4. Preparation in 80-cm² Flasks

1. Decant medium from flasks.
2. Add 2 mL of virus dilution in TBS at 1–3 FFU/cell.
3. Incubate for 60–90 min at 37°C.
4. Add 50 mL medium +0.5% FCS.
5. Incubate for 3–5 d.
6. When good CPE is visible, harvest virus by shaking cells off into medium, spinning down, and proceeding as above (**Subheading 3.3., steps 7–11**).

3.5. Rapid Preparation of Banded Virus (11)

1. Make large quantities of virus stock (e.g., from 30 60-mm plates or 10 80-mL flasks).
2. Add 1/100 vol *n*-butanol and incubate at 4°C for 60 min.
3. Centrifuge at 900g for 10 min to remove cell debris.
4. Remove supernatant containing virus.
5. Layer on a CsCl/glycerol gradient (*see* **Subheading 3.6.**).
6. Centrifuge at 80,000g for 90 min at 4°C. Do not use brake to stop.
7. Virus band should be visible to naked eye as the lowest opalescent band. To see more clearly, place in front of a black background and shine a light down through

the tube from directly above. There may be one or more bands above the virus band; these consist of empty or incomplete virus particles.

8. Collect the virus band by piercing the tube with a hypodermic needle below the band, and collecting drops into sterile 1-mL cryotubes. The fraction containing the virus band will be opalescent.
9. Remove CsCl either by dialysis against 10 mM Tris-HCl, 1 mM EDTA, pH 8.0, or PBS/10% glycerol, or on a G50 Sephadex desalting column (*see* **Subheading 3.8.**).
10. Mix virus with an equal volume of sterile 80% glycerol and store at –70°C.

3.6. CsCl Gradient

Use Beckman 14 × 95-mm Ultraclear tubes (Fullerton, CA, cat. no. 344060) in an SW40Ti swing-out rotor. Prepare the gradient by layering the following solutions; it is easier to put the glycerol in first, then underlay it with the ρ = 1.32 CsCl, underlay this with the ρ = 1.45 CsCl, and finally to overlay the virus. Use 2 mL of ρ = 1.45 CsCl, 3 mL of ρ = 1.32 CsCl, 2 mL of 40% glycerol, and approx 7 mL of virus prep. If necessary, fill up tube with 10 mM Tris-HCl, 1 mM EDTA, pH 8.0. Centrifuge at 80,000g for 90 min at 4°C.

3.7. Equilibrium Banding of Virus

1. Store the virus band from the rapid purification at 4°C without desalting (prolonged storage is not recommended).
2. Centrifuge on a CsCl step gradient at 85,000g overnight at 4°C in a 5-mL swing-out rotor. CsCl gradient: Use 1.5 mL of ρ = 1.45, 1.8 mL of ρ = 1.32, and 2.5–3 mL of virus.
3. Add a thin layer of liquid paraffin to seal.
4. Collect virus band via a hypodermic needle in 10-drop fractions.
5. Store at 4°C, or desalt, add glycerol to 40%, and store at –70°C.

3.8. Storage of Banded Virus

Banded virus taken directly from a CsCl gradient can be kept for a few days at 4°C, but there is a danger that the virus will aggregate and infectivity will be lost. The virus will remain in suspension for prolonged periods (months or years) if glycerol is added to 40% and the material is aliquoted and stored at –70°C. Such stock can be used for infecting cells after dilution in TBS. To determine the virus concentration, it is a good idea to take an aliquot before adding glycerol.

CsCl can be removed by dialysis or desalting: the virus will be diluted some two- to fivefold, and there is a risk that the virus will aggregate during these procedures. The method of choice for desalting will be determined by the purpose for which the material is intended. For particle counts and other assays of virus concentration, virus can be desalted but glycerol should not be added. Virus intended for subsequent infection can be stored at –70°C with addition of glycerol after desalting.

1. Desalt on a G50 Sephadex column preequilibrated with TBS; or
2. Dialyze vs TBS *or* 60 m*M* Tris-HCl, pH 7.5, 10 m*M* EDTA, 2% glycerol *or* vs 10 m*M* Tris-HCl, pH 7.6, 1 m*M* MgCl$_2$, 10% glycerol, 0.5% *n*-butanol *(12)*; and/or
3. Mix virus with an equal volume of sterile 80% glycerol and store at –70°C.

3.9. Preparation of Virion DNA From Banded Virus

1. Use virus directly from the rapid-banding stage (*see* **Subheading 3.5.**).
2. Add Proteinase K to 500 mg/mL and SDS to 0.5% (final concentration).
3. Incubate at 37°C for 2–3 h.
4. Extract DNA once with phenol/chloroform (equilibrated in T/E), then once with chloroform (equilibrated in T/E).
5. Ethanol-precipitate with 0.3 *M* sodium acetate, pH 5.5, at –70°C for 20 min, or overnight at –20°C.
6. Wash with 70% ethanol at –20°C.
7. Dissolve pellet in T/E.

3.10. Preparation of Viral DNA From Infected Cells (Modified Hirt Extraction) (13)

1. For 60-mm plate:
   a. If the cells are firmly adhering, wash in TBS, drain, add 0.6 mL of Hirt buffer, incubate for 10 min at room temperature, and scrape lysate into Eppendorf tube.
   b. If cells are detaching from plate, scrape cells into medium, centrifuge at 2000*g* for 10 min or at 15,000*g* for 1 min, resuspend cell pellet in 0.6 mL Hirt buffer, transfer to Eppendorf tube, and incubate for 10 min at room temperature.
2. Add 0.15 mL 5 *M* NaCl; mix gently by inverting tube.
3. Incubate on ice overnight.
4. Centrifuge at 15,000*g* for 30–45 min.
5. Remove supernatant (sometimes it is easier to remove pellet with a wooden toothpick).
6. Precipitate DNA from supernatant using either of two methods:
   a. Add equal volume isopropanol, incubate –20°C overnight, centrifuge at 12,000*g* for 5 min, and drain pellet.
   b. Add Proteinase K to final concentration 500 μg/mL, incubate 37°C for 3 h, extract once with phenol/chloroform, extract once with chloroform, ethanol-precipitate with 0.3 *M* sodium acetate, and wash pellet in 70% ethanol.

3.11. Preparation of Virion Packaged DNA From Infected Cells (14)

This method is useful for preparing DNA from a seed stock to assess the quality and quantity of the yield by DNA restriction-digest analysis. For 150 μL seed stock, proceed directly to **step 7**. For two 60-mm plates of infected cells:

1. Scrape cells into medium and centrifuge to pellet cells at 900*g* for 10 min or 15,000*g* for 1 min.

2. Wash cell pellet once in TBS.
3. To drained pellet, add 0.5 mL TBS.
4. Freeze–thaw three times.
5. Centrifuge at 15,000g for 1 min.
6. Collect supernatant.
7. To supernatant, add Proteinase K to a final concentration of 500 µg/mL.
8. Incubate at 37°C for 2–3 h.
9. Extract once with phenol/chloroform; extract once with chloroform.
10. Ethanol-precipitate with 0.3 M sodium acetate.

For smaller samples (e.g., from Linbro well or 35-mm plate), add salmon sperm DNA or tRNA as carrier for DNA precipitation.

3.12. Fluorescent Focus Assay

1. Set up monolayers of cells in 35-mm plates or Linbro wells at 3×10^4 cells/well.
2. Wash with complete PBS.
3. Infect with 0.1 mL virus dilution in TBS.
4. Adsorb virus for 90 min at 37°C, shaking plates every 20–25 min.
5. Overlay with DMEM + 0.5% FCS.
6. Incubate for 24 or 48 h.
7. Remove medium; wash twice with complete PBS. Inspect cells to confirm morphology and adherence.
8. Fix with ice-cold 90% methanol for 4 min.
9. Wash twice with PBS. Plates may now be stored at 4°C with 1 mL of PBS overlay.
10. Add antibody at dilutions 1:50, 1:150, 1:450 in PBS at 0.4 mL/plate. Can use group-specific mouse monoclonal antibody such as 10/5.1.2 NCL-Adeno (Novocastra, Newcastle, UK).
11. As a control, use preimmune serum where available. Additional control: Omit first antibody.
12. Incubate at room temperature for 30 min.
13. Save antiserum and store at 4°C; can be reused at least twice.
14. Wash plates twice in PBS.
15. Add 0.4 mL per plate fluorescein-conjugated goat anti-mouse IgG and rhodamine-conjugated BSA at appropriate dilution; this needs to be determined empirically. Goat anti-mouse serum can be reused at least twice.
16. Shake every 10 min at room temperature for 30 min.
17. Rinse twice in PBS. Plates can be stored in dark at 4°C with 1 mL PBS.
18. Read plates with UV microscope ×10 objective, ×0 eyepiece with 1 mm grating graticule.
19. Count the total cell number under phase-contrast optics, and the number of fluorescing cells in the same field.

3.13. Virus Concentration

These virus concentration methods were *(15–19)* devised principally for the subgroup C viruses Ad2 and Ad5, but Ad40 is considered to be sufficiently similar so that the methods remain valid. It should be noted that there are discrepancies between the values obtained by the various methods, and it is recommended that for consistency, one method should be chosen and adhered to.

3.13.1. DNA Determination

1. Using the PicoGreen dsDNA Quantitation Kit™ (Invitrogen/Molecular Probes) dilute the stock T/E (from the kit) 1:20 in H_2O.
2. Prepare a standard curve of lambda bacteriophage dsDNA in the range 0.01–1.0 µg/mL, 100-µL aliquots in the dilute T/E.
3. Dilute 10- and 3-µL aliquots of banded virus to 100 µL with dilute T/E.
4. Heat-inactivate virus aliquots at 56°C for 30 min.
5. Dilute stock PicoGreen 1:200 in T/E (use plastic not glass container and keep in dark).
6. In a 96-well clear plate mix 100 µL virus sample or DNA standard plus 1 µL 10% SDS.
7. Incubate at room temperature (RT) for 5 min.
8. Add 100 µL diluted PicoGreen reagent.
9. Incubate at room temperature for 2 min in dark.
10. Read on fluorescence plate reader set at 485 nm excitation and 535 nm emission (fluorescein settings).
11. Calculate DNA concentration by linear regression from standard curve.
12. For Ad40, genome = 34214bp, and 1 µg dsDNA = 2.67×10^{10} particles.

3.13.2. Absorbance at A_{260nm}

1. Using banded virus, dilute 1:10 into 0.5% SDS, 0.1X SSC. 1 OD_{260} of virus in 0.5% SDS, 0.1X SSC = 0.28 mg/mL protein. This is equal to 11×10^{11} virus particles.
2. Extract viral DNA and measure OD_{260}. 1 OD_{260} = 10^{12} virus particles/mL.

3.13.3. Protein Determination

10 µL CsCl banded virus + 100 µL H_2O + 200 µL Bio-Rad protein reagent (undiluted). 1 OD_{595} = 3.4×10^{12} virus particles.

3.13.4. Virus Particle Count in the Electron Microscope

This method is from Jim Aitken, Institute of Virology, Glasgow. Virus must be desalted.

1. Mix sodium silicotungstate or phosphotungstic acid stain (**Heading 2.**, **item 24**), virus suspension, and latex beads (**Heading 2.**, **item 21**) at room temperature.
2. Put a few µL onto an electron microscopic grid (can be formvar, collodion, or carbon coated; 200, 300, or 400 mesh size).

3. Allow to incubate for a few minutes. Drain off excess liquid by lightly drawing Whatman no. 1 filter paper across the grid surface.
4. Count number of virus particles and simultaneously count latex beads. If there is a large number of empty or incomplete particles (penetrated by stain), it is worth recording this number also.
5. When you get to 100 beads, stop counting.
6. Particle count = $n/100 \times$ concentration of latex beads (where n is number of virus particles counted).

4. Notes

1. Safety considerations for working with Ad40: Ad40 is generally considered to be nonpathogenic to adults and is characteristically associated with infantile gastro-enteritis *(2)*.
2. The virus is classified as nontumorigenic, and in fact only partial transformation can be achieved in vitro *(20)*. However, given that the virus is infectious by the oral route, it is prudent to confine work with large amounts of purified virus to a designated and contained area. The extreme resistance of Ad40 to UV irradiation has been reported *(21)*. The actual level of containment will be governed by local regulations. The use of gloves and an appropriate side- or back-fastening lab coat is recommended.
3. Manipulations of virus-infected material should be designed to minimize the generation of aerosols; if material is vigorously shaken or vortexed, the tube should be briefly centrifuged prior to opening. Waste materials should be rendered non-infectious prior to disposal: 1% Virkon or 0.5% SDS is sufficient to inactivate virus, but it is important to ensure that solid waste is either autoclaved or exposed to disinfectant for a sufficient length of time (at least 30 min) before discarding.
4. The use of glassware and hypodermic needles should be kept to a minimum; any skin contact or needlestick injury should be thoroughly flushed with running water. Virkon is suitable for external skin application. A drip tray should be used to contain spillages, which can be absorbed onto paper tissue soaked in disinfectant.

Acknowledgments

This chapter is dedicated to the memory of Helio Pereira and Harry Ginsberg, two of the founding fathers of adenovirology. I am indebted to Gaie Brown, Nancy Mackay, and Angela Rinaldi, who helped to develop many of the methods described here, and whose expert technical assistance over many years is much appreciated.

References

1. Mautner, V., Steinthorsdottir, V., and Bailey, A. (1995) The enteric adenoviruses, in *The Molecular Repertoire of Adenoviruses* (Doerfler. W. and Boehm P., eds.), *Curr. Topics Microbiol. Immunol.* **199/III**, Springer Verlag, New York, pp. 229–282.

2. Favier, A.L., Schoehn, G., Jaquinod, M., Harsi, C., and Chroboczek, J. (2002) Structural studies of human enteric adenovirus type 41. *Virology* **293,** 75–85.

3. deJong, J. C., Wigand, R., Kidd, A. H., et al. (1983) Candidate adenoviruses 40 and 41: fastidious adenoviruses from human infant stool. *J. Med. Virol.* **11,** 215–231.

4. Babiss, L. E., Young, C. S. H., Fisher, P. B., and Ginsberg, H. S. (1983) Expression of adenovirus EIA and EIB gene products and the *Escherichia coli* XGPRT gene in KB cells. *J. Virol.* **46,** 454–465.

5. Davison, A. J., Telford, E. A. R., Watson, M. S., McBride, K., and Mautner, V. (1993) The DNA sequence of adenovirus type 40. *J. Mol. Biol.* **234,** 1308–1316.

6. Adrian, T., Wadell, G., Hierholzer, J. C., and Wigand, R. (1986) DNA restriction analysis of adenovirus prototypes 1 to 41. *Arch. Virol.* **91,** 277–290.

7. Hashimoto, S., Sakakibara, N., Kumai, H., et al. (1991) Fastidious adenovirus type 40 can propagate efficiently and produce plaques on a human cell line, A549, derived from lung carcinoma. *J. Virol.* **65,** 2429–2435.

8. Brown, M., Wilson-Friesen, H. L., and Doane, F. (1992) A block in release of progeny virus and a high particle-to-infectious unit ratio contribute to poor growth of enteric adenovirus types 40 and 41 in cell culture. *J. Virol.* **66,** 3198–3205.

9. Tiemessen, C. T. and Kidd, A. H. (1994) Adenovirus type 40 and 41 growth in vitro: host range diversity reflected by differences in patterns of DNA replication. *J. Virol.* **68,** 1239–1244.

10. Graham, F. L., Smiley, J., Russell, W. C., and Nairn, R. (1977) Characteristics of a human cell line transformed by DNA from human adenovirus type 5. *J. Gen. Virol.* **36,** 59–72.

11. Mautner, V. and Willcox, H. N. A. (1974) Adenovirus antigens: a model system in mice for sub-unit vaccination. *J. Gen. Virol.* **25,** 325–336.

12. Hannan, C., Raptis, L. H., Dery, C. V., and Weber, J. (1983) Biological and structural studies with an adenovirus type 2 temperature-sensitive mutant defective for uncoating. *Intervirology* **19,** 213–223.

13. Hirt, B. (1967) Selective extraction of polyoma DNA from infected mouse cell cultures. *J. Mol. Biol.* **26,** 365–369.

14. Mautner, V., Mackay, N., and Steinthorsdottir, V. (1989) Complementation of enteric adenovirus type 40 for lytic growth in tissue culture by EIB 55K function of adenovirus types 5 and 12. *Virology* **171,** 619–622.

15. Green, M. and Pina, M. (1963) Biochemical studies on adenovirus multiplication. IV. Isolation and purification and chemical analysis of adenovirus. *Virology* **20,** 199–207.

16. Chardonnet, Y. and Dales, S. (1970) Early events in the interaction between adenoviruses and HeLa cells. II. Penetration of type 5 and intracellular release of the DNA genome. *Virology* **40,** 462–477.

17. Laver, W. G. (1970) Isolation of an arginine-rich protein from particles of adenovirus type 2. *Virology* **41,** 488–500.

18. Russell, W. C., McIntosh, K., and Skehel, J. J. (1971) The preparation and properties of adenovirus cores. *J. Gen. Virol.* **11,** 35–46.

19. Murakami, P. and McCaman, M. T. (1999) Quantitation of adenovirus DNA and virus particles with the PicoGreen fluorescent dye. *Anal. Biochem.* **274,** 283–288.
20. van Loon, A. E., Maas, R., Vaessen, R. T., Reemst, A. M., Sussenbach, J. S., and Rozijn, T. H. (1985) Cell transformation by the left terminal regions of the adenovirus 40 and 41 genomes. *Virology* **147,** 227–230.
21. Thurston-Enriquez, J. A., Haas, C. N., Jacangelo, J., Riley, K., and Gerba, C. P. (2003) Inactivation of feline calicivirus and adenovirus type 40 by UV radiation. *Appl. Environ. Microbiol.* **69,** 577–582.

12

Immunocompetent, Semi-Permissive Cotton Rat Tumor Model for the Evaluation of Oncolytic Adenoviruses

Karoly Toth, Jacqueline F. Spencer, and William S. M. Wold

Summary

Oncolytic adenovirus (Ad) vectors belong to a new class of cancer therapy agents that destroy cancer cells as part of the virus's lytic infectious cycle. In this chapter we describe an immunocompetent, semi-permissive cotton rat tumor model to evaluate the safety and efficacy of oncolytic Ad vectors. With this model one can investigate the effect of the host immune system on the vector–tumor interaction as well as the vector's effect on normal host cells in vivo. This chapter describes procedures for analyzing the growth and cytolytic properties of oncolytic Ad vectors in cotton rat cells in vitro. We discuss handling and husbandry issues and techniques for subcutaneous, intratumoral, and intravenous injection of cotton rats. We present methods for generating subcutaneous tumors in cotton rats and assessing the efficacy of Ad vectors upon intratumoral injection. Also, we discuss procedures for determining the biodistribution of a replicating Ad in cotton rats.

Key Words: Adenovirus; tumor; oncolytic; cotton rat; immunocompetent; permissive; replicating.

1. Introduction

Oncolytic adenovirus (Ad) vectors belong to a new class of cancer therapy agents. These vectors infect and destroy cancer cells as part of the virus's lytic infectious cycle. To evaluate the safety and efficacy of oncolytic Ad vectors, a satisfactory animal model is needed. In order to assess adequately all facets of the interactions of the vector, the tumor, and the host animal, this animal model should be immunocompetent and permissive for the replication of Ad serotype 5 (Ad5) (most oncolytic Ads constructed to date are based on Ad5). The animal should also support the growth of implanted tumors in which human Ads replicate, and mimic human disease induced by replicating Ads. Unfortunately,

From: *Methods in Molecular Medicine, Vol. 130:*
Adenovirus Methods and Protocols, Second Edition, vol. 1:
Adenoviruses Ad Vectors, Quantitation, and Animal Models
Edited by: W. S. M. Wold and A. E. Tollefson © Humana Press Inc., Totowa, NJ

most conventionally used laboratory rodents are refractive for Ad replication. Until recently, Ad vectors have been evaluated primarily in immunodeficient mice bearing human tumor xenografts. However, this model cannot satisfactorily address the effect of the host immune system on the vector-infected tumor or the toxicity of the vector in normal tissues. To resolve this inadequacy we have developed a cotton rat tumor model. Upon infection, Ad5 has been shown to replicate in the lungs, corneas, and conjunctiva of cotton rats and cause an illness akin to that seen in Ad5-infected humans *(1–5)*. For these reasons, cotton rats are ideally suited to study the interaction of the host, the tumor, and the oncolytic Ad vector in its full complexity.

This chapter describes methods for studying Ad infection of a cotton rat cell line and generating subcutaneous tumors in cotton rats using this cell line. Also, we discuss methods for assessing the efficacy of an oncolytic Ad in subcutaneous tumors in cotton rats and to study the biodistribution of the vector after intratumoral injection. Furthermore, recommendations for general handling and husbandry of cotton rats will be given.

2. Materials

1. Cotton rats (*Sigmodon hispidus*) can be purchased from Harlan Sprague Dawley (HSD, Indianapolis, IN). In our experiments, we used 4- to 5-wk-old female animals. The LCRT cell line was obtained from Virion Systems (Rockville, MD).
2. LCRT cells were derived from a breast tumor with sarcomatous and carcinomatous elements; upon passaging only the sarcomatous element (fibrosarcoma) remained. The cell lineage has not been cloned *(6)*.
3. Phosphate-buffered saline (PBS) with bivalent cations (PBS^{++}): powdered PBS can be purchased from Sigma (St. Louis, MO). After dissolving and adjusting the pH according to the manufacturer's recommendations, add 1 mM CaCl$_2$ and 1 mM MgCl$_2$. Make it up to the desired volume and filter-sterilize.
4. Dulbecco's modified Eagle's medium (DMEM).
5. Trypsin–ethylene diamine tetraacetic acid (EDTA) solution.
6. Fetal bovine serum (FBS).
7. Crystal violet stain is 1% crystal violet, 20% ethanol, and 10% formaldehyde in water.
8. Tissue culture plasticware: T-75 flasks, T-450 flasks (TPP AG, Trasadingen, Switzerland).
9. Polycarbonate rat cages, containing a stainless steel wire bar lid that snaps down tightly (Allentown Caging, Allentown, NJ).
10. Tekfresh rodent bedding.
11. 10-cm Piece of 3" PVC pipe.
12. Standard rodent diet containing at least 18% protein (HSD Teklad 2018).
13. Automated water system or water bottles fitted with sipper tubes.
14. Kevlar gloves with latex-coated palms (American Health and Safety, Madison, WI).
15. CO$_2$/O$_2$ (USP 80/20 mix).

16. An anesthetic cocktail containing 30 mg/mL of ketamine and 10 mg/mL of xylazine in water for injection.
17. Tuberculin syringe with a 25G × 5/8" needle (Becton Dickinson, Franklin Lakes, NJ).
18. Puralube petrolatum ophthalmic ointment (Fougera, Melville, NY).
19. Circulating water blanket (Gaymar, Orchard Park, NY).
20. Wahl Peanut clippers (Wahl, Sterling, IL) or other appropriate clippers.
21. 70% Ethanol.
22. 0.3-mL Insulin syringe with an attached 29G × 0.5" needle (Terumo, Elkton, MD).
23. Rodent work stand (Braintree Scientific, Braintree, MA).
24. Small roll of gauze.
25. Micropore surgical tape (3M, St. Paul, MN).
26. Bio-Pulverizer device (Research Products International, Mount Prospect, IL) or bead-beater type homogenizer like the Qiagen TissueLyser (Qiagen, Valencia, CA), depending on number of samples (*see* **Note 8**).

3. Methods
3.1. Culturing LCRT Cells

1. LCRT cells can be cultured as an adherent monolayer (*see* **Note 1**) in DMEM containing L-glutamine and 10% FBS at 37°C in a humidified incubator with 5% CO_2 content.
2. Passage the cells every 3 d (when cells reach about 80–90% confluency) at a dilution of 1:3 to 1:6. Remove the medium from the monolayer, and wash it with prewarmed (37°C) PBS. Add prewarmed trypsin-EDTA solution (1 mL for a T-75 flask), and incubate it at 37°C for 5 min (or until all cells are rounded up). Collect the cells by adding 5 mL of DMEM with 10% FBS and gently pipetting the medium up and down. Use the appropriate volume of this cell suspension to seed another tissue culture vessel; add the required volume of fresh medium.
3. For large-scale culture (e.g., producing cells for inducing subcutaneous tumors), we routinely use T-450 flasks. These vessels contain three horizontal shelves that provide a growing area of 450 cm² and yield approx 3×10^7 cells (grown to 90% confluency).
   a. Seed the flasks with about 5×10^6 LCRT cells in 80 mL of DMEM with 10% FBS, and let them grow to the required density.
   b. Wash the cells with 100 mL PBS. The PBS can be "reused" for washing subsequent flasks; an experienced person can process six flasks using the same solution. Discard the PBS after washing the last flask. Speed is important as cells may dry out if left without any liquid on them for a long time.
   c. Add 50 mL of trypsin–EDTA to the first flask, ensure that all the shelves are completely wetted, and transfer the trypsin–EDTA solution to the next flask. Lay the first flask horizontally on the bench. Again, six flasks can be trypsinized using the same solution. Check the cells under a phase-contrast microscope. By the time the sixth flask has had the trypsin–EDTA treatment, cells may be collected from the flask trypsinized first.

d. Collect the cells by adding 100 mL of DMEM with 10% FBS to the flask. Agitate the flask gently to dislodge attached cells. When all cells are floating, transfer the medium to the second flask, and sequentially collect cells from all flasks.

3.2. Assess the Replication of an Oncolytic Ad in LCRT Cells

The replication capabilities of an oncolytic Ad can be ascertained by a single-cycle replication assay. In this assay cells are infected at a multiplicity of infection (MOI) that ensures the simultaneous infection of all cells. Infected cells and medium are collected at several time points and tested for progeny virus. By performing this experiment using a wild-type (wt) Ad as the positive control, one can estimate if the tested oncolytic Ad will be replicating to a sufficient degree in LCRT cells (*6*).

1. Prepare a number of 35-mm tissue culture dishes containing LCRT cells at about 60–70% confluency. The number of dishes needed can be calculated by multiplying the number of Ads to be tested by the number of planned harvest times (*see* **Note 2**) and adding one dish for counting the cells.
2. Count the number of cells on one dish by trypsinizing all the cells and counting them in a hemacytometer. Discard this dish.
3. Infect parallel dishes (one for each harvest time point planned) with 50 plaque-forming units (PFU)/cell of virus according to procedures described in Chapter 18. Repeat this for each virus.
4. After the incubation period, wash all dishes three times with serum-free DMEM. Aspirate the last wash, and add 2 mL of DMEM containing 5% FBS. Incubate the dishes in a tissue culture incubator (*see* **Note 3**).
5. Harvest one dish for each virus at the defined time points by freezing it in a –80°C freezer. It is important to collect the cells *and* the medium together because Ads are released into the medium from lysed cells. Samples can be stored frozen at –80°C.
6. To release virions from cells, freeze–thaw cells three times (transfer the samples into 5-mL snap-cap centrifuge tubes, be sure that all the cells and the medium are transferred), and sonicate the samples as described in Chapter 18.
7. Clarify the crude lysates by pelleting the cell debris in a tabletop centrifuge at about 250*g*. Transfer the supernatants into fresh tubes, and discard the pellets.
8. Plaque-titer the samples on A549 cells as described in Chapter 18.

3.3. Investigate the Ability of an Oncolytic Ad to Spread Cell-to-Cell in LCRT Cells

Oncolytic Ads are expected to produce progeny virus and infect neighboring cells in the infected tumor. To assess the speed of this infection–virus production–release–reinfection cycle, a virus-spread assay can be performed. In this assay, parallel vessels containing the cell line are infected with various multiplicities (generally ranging from 10^2 to 10^{-4} PFU/cell) of an oncolytic Ad. The cells infected with high MOI will be lysed within one infectious cycle, whereas multiple virus release–infection cycles are needed to produce cytopathic effect (CPE) in vessels infected with low MOI. By recording the extent of CPE caused by Ad infection at different MOI at a given time and comparing it to CPE caused by a wt Ad, one can judge the ability of an oncolytic Ad to spread cell-to-cell *(7–10)*.

1. Prepare a 48-well tissue culture plate (*see* **Note 4**) containing LCRT cells at about 60–70% confluency.
2. Count the cells in one well by trypsinizing all the cells and counting them in a hemacytometer. (Count a well in the mock-infected control column.)
3. Calculate 10 times the amount of virus (for each test virus and control) needed for the highest multiplicity planned. Suspend it in 1 mL of serum-free DMEM in a 5-mL snap-cap centrifuge tube.
4. Make serial dilutions starting with adding 100 µL of the virus suspension to 900 µL of serum-free DMEM. Repeat this procedure until six serial 10-fold dilutions are made. (For a detailed description of serial dilution of virus stocks, Chapter 18.)
5. Aspirate the medium from the 48-well plate (*see* **Note 5**).
6. Add 100 µL of virus suspension to each well using an automatic pipettor, starting from the most dilute and working toward the most concentrated (there is no need to change the pipet tip for a single virus series). Repeat it for every test and control virus.
7. Incubate the plate for 1 h in a tissue-culture incubator.
8. Add 100 µL of DMEM containing 10% FBS to each well.
9. Incubate for 6–20 d. If medium levels drop because of evaporation (*see* **Note 3**), replenish the level with serum-free DMEM. *Do not change the medium.*
10. Observe the plate through phase-contrast microscopy periodically. Stain the cells when CPE develops in the well infected with the lowest MOI with one of the vectors.
11. Aspirate the medium from all the wells (*see* **Note 6**). Stain the remaining monolayer with crystal violet solution for 10 min. Aspirate the staining solution. Wash the plate with tap water by holding it parallel to the stream of water and flicking the water out when the wells become filled up; repeat this procedure until there is no stain left on the clear plastic. Air-dry the plate. The blue stain indicates intact monolayer (cells attached to the plastic); the transparent areas are spots where the virus caused lysis/detachment of cells (**Fig. 1**).
12. A rough estimation of the cells remaining attached to the plastic surface can be performed by adding 200 µL of 30% acetic acid to the wells. This will dissolve the crystal violet dye, which can then be quantitated by measuring the absorbance of the solution at 560 nm. The OD_{560} value will be inversely proportional to the viral CPE.

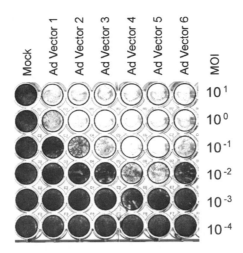

Fig. 1. Result of a "spread assay." Cells were infected with serial dilutions of six oncolytic Ad vectors, ranging from 10 to 10^{-4} PFU/cell. The infections were arranged in parallel columns of a 48-well plate to facilitate comparisons between the vectors. Viral cytopathic effect was visualized by crystal violet staining.

3.4. Handling and Husbandry of Cotton Rats

1. Caging: cotton rats are extremely excitable animals that are difficult to handle. For the convenience of the handler, cotton rats should be single-housed in polycarbonate rat cages, containing a stainless steel wire bar lid that snaps down tightly. Tekfresh rodent bedding is placed in the bottom of the cage for nesting. For environmental enrichment, a 10-cm piece of 3" PVC pipe is placed in the cage. This pipe gives the animal a place to hide and helps to decrease the animal's stress.

2. Diet: the animals should be fed a standard rodent diet containing at least 18% protein (HSD Teklad 2018) and have free access to water. Because the cotton rats will likely experience shipping stress, food and water consumption should be closely monitored during acclimation. An acclimation period of at least 7 d is recommended.

3. Physical restraint: cotton rats have a strong fight-or-flight response and tend to bite when being caught and handled. During routine cage changing, it is recommended that the animal handler wear Kevlar gloves with latex-coated palms or some other type of thick glove for protection. The animal should not be handled by the tail, as degloving of the tail will occur. When removing the animal from the cage, the lid should be opened just enough for the handler to reach his/her hand into the cage. This will help to prevent the animal from escaping *(11)*. The PVC pipe can be used to trap the animal for transfer into its new cage. The animal may be physically restrained by scruffing the skin over the shoulders and back tightly, while pushing the animal down with an appropriate amount of force.

Pushing the animal down too firmly may result in the cotton rat biting its tongue and bleeding from the mouth.

4. Short-term anesthetic restraint: cotton rats appear very stressed during catch-and-restraint procedures. CO_2/O_2 (USP 80/20 mix) can be used to lightly anesthetize the animals and facilitate physical restraint. This can be accomplished by pumping the CO_2/O_2 mixture into the animal's cage via a small tube until the animal loses consciousness. CO_2 gas is heavier than air and tends to persist inside the cage for a long period of time; therefore, after the restraint procedure, the animal should be either placed into a new cage or returned to the original cage after the CO_2 gas has dissipated or is fanned out.

5. Long-term anesthetic restraint: for certain procedures it is necessary to have the animal immobilized for an extended period. An anesthetic cocktail containing 30 mg/mL of ketamine and 10 mg/mL of xylazine in water for injection can be given at 0.1 mL/100 g body weight via intraperitoneal injection. The cotton rat should be scruff-restrained, as described above, and placed in dorsal recumbency with the head pointing downward for injection. A tuberculin syringe with a 25G × 5/8″ needle is used for injecting the anesthetic cocktail. The needle is inserted into the right lower quadrant of the abdomen at approximately a 45° angle. The syringe plunger should be pulled back slightly to ensure correct needle placement before the anesthetic is injected. If blood, feces, or urine is aspirated into the syringe, remove and discard the syringe and repeat the procedure using a new syringe of anesthetic cocktail. After injection, the animal should be placed into a cage without bedding until it loses consciousness. When the animal is fully anesthetized, Puralube petrolatum ophthalmic ointment must be placed in the animal's eyes to prevent corneal dryness and abrasions. During anesthetic recovery, place the cotton rat in the cage without bedding, and set it on a prewarmed circulating water blanket. An anesthetized animal's airway can become blocked if it aspirates bedding; therefore, cages without bedding are used for anesthesia and recovery. The animal can be returned to its original cage when it is ambulatory (the animal is mobile and has righting reflex).

3.5. Injecting Cotton Rats

3.5.1. Subcutaneous Injection

Subcutaneous injection of cotton rats generally requires two people: one to restrain the animal and one to perform the injection. The cotton rat can be restrained using the methods described above. Short-term anesthetic restraint may be used to facilitate animal handling. The injection site(s) should be shaved using Wahl Peanut clippers or other appropriate clippers. The clipper operator should take care not to nick the skin. The shaved site should be swabbed with 70% ethanol prior to injection. A tuberculin syringe with a 25G × 5/8″ needle is used for the injection. The skin at the injection site should be lifted slightly for placement of the needle. The needle is inserted bevel side up under the skin and then the contents of the syringe injected. The correct placement of the needle can be checked by slightly pulling back on the plunger (this ensures it is not in a blood vessel), and the needle should move freely under the skin. The contents of the syringe should evacuate with minimal pressure exerted on the plunger. If the syringe contents feel hard to inject, the needle may be incorrectly placed either intradermally or intramuscularly. After the injection, a small raised area may be present at the site.

3.5.2. Intratumoral Injection

The injection site should be shaved as described above. A 0.3-mL insulin syringe with an attached 29G × 0.5" needle is used for injection. The animal should be scruff-restrained as described earlier. The needle should be inserted as deep as possible into the tumor with the bevel side up. During injection, the needle is repositioned in a fan-like pattern as it is being withdrawn. After all of the syringe contents are evacuated and before completely withdrawing the needle, the syringe is twisted clockwise and withdrawn in a downward motion out of the tumor. This helps prevent leakage of the injectable out of the tumor.

3.5.3. Intravenous Injection

Standard methods for intravenous injection in other rodent species do not work well in cotton rats. The tail veins are inaccessible because of the tail's coloration and the tendency for degloving. Cardiac puncture is effective for blood collection and injection, but is contraindicated because of the risk of lung puncture and exsanguination *(12)*. The jugular vein is the preferred route for intravenous injection in the cotton rat. The vessel is easily accessible, and using this route of administration poses little risk of accidental death or injury to the animal. The cotton rat should be anesthetized using the long-term anesthetic protocol described above (**Subheading 3.4.5.**). When the animal is almost fully anesthetized, shave the ventral neck area with Wahl Peanut hair clippers from below the clavicles to midway on the neck. When the cotton rat is fully anesthetized, the injection site is swabbed with 70% ethanol. The animal should then be restrained on a rodent work stand. Loop the upper front teeth with the tooth loop (provided with the stand) and position the animal on the stand in dorsal recumbency. Place a small roll of gauze (approx 1/2" in diameter) under the base of the cotton rat's neck (dorsal side) so that the head is slightly hyperextended. The jugular vein is located approximately at the midpoint of the clavicle. To facilitate access to the vessel, use Micropore surgical tape to tape down the front legs to the side arms of the stand. If the clavicle and the jugular vein cannot be visualized, it may be necessary to palpate the clavi-cle to determine the midpoint. A 0.3-mL insulin syringe with an attached 29G × 0.5" needle (Terumo) is used for injection. Position the syringe at a 45° angle just cranial to the clavicle. Enter the skin with the needle bevel up, and pull back on the plunger gently as the needle is advanced. As the needle enters the vessel, blood will flow into the barrel of the syringe. If the needle has been advanced more than three-quarters of the way and no blood has entered the syringe, the needle has missed or advanced through the vein. Keep slight backpressure on the plunger, and slowly back the needle out of the animal. The needle may enter the vessel during withdrawal of the syringe. Once blood is flowing into the syringe barrel, the syringe contents may be slowly injected. The injector must keep the syringe position stable. Any repositioning of the needle may result in perivascular injection. If the syringe plunger becomes harder to advance or a raised area becomes visible at the injection site, the syringe may need to be repositioned and the patentcy of the needle checked. After injecting, remove the needle and hold off the vessel using mild pressure for approx 5–10 s. Gently remove the animal from the stand and remove the tooth loop. Recover the animal from anesthesia as described above (**Subheading 3.4.5.**).

3.6. Generate Subcutaneous LCRT Tumors in Cotton Rats

LCRT cells form fast-growing, invasive tumors when injected subcutaneously into cotton rats. These tumors have a large necrotic/hemorrhagic center. They often metastasize to the draining lymph nodes and to the lung. The antitumor efficacy of an oncolytic Ad can be estimated by recording the growth of the subcutaneous tumor after vector administration *(6–8,10,13)*.

1. Grow a large-scale culture of LCRT cells (*see* **Subheading 3.1.3.**); harvest and count cells in a hemacytometer.
2. Pellet the cells in a tabletop centrifuge at 250*g*. Resuspend the cell pellet in serum-free DMEM to yield 1×10^7 cells/mL. Keep the suspension on ice.
3. Inject 100 µL (1×10^6 cells) subcutaneously into both hind flanks of a cotton rat (*see* **Subheading 3.5.1.**) Approximately 100- to 400-µL tumors will be apparent in about 3 d. The tumors will grow to 5–10 mL in 14 d.
4. Inject the oncolytic Ad to be tested intratumorally or intravenously (*see* **Subheadings 3.5.2.** and **3.5.3.**), and follow the growth of the tumor by measuring it at intervals (twice weekly is adequate) with a digital caliper.

3.7. Assaying the Biodistribution of Intratumorally Injected Ads in Cotton Rats

Because cotton rats are semi-permissive for human Ads, the distribution of an Ad after intratumoral injection is expected to resemble the distribution of this virus in cancer patients treated with Ad vectors. These experiments can give valuable clues as to which organs might be secondary targets of a given oncolytic Ad when it is injected intratumorally *(6)*.

1. Generate subcutaneous LCRT tumors in cotton rats (*see* **Subheading 3.6.**).
2. When the tumors reach approx 300–500 µL, inject them with the test Ad (*see* **Subheading 3.5.2.**).
3. Sacrifice the animals by CO_2 suffocation at the planned time points (*see* **Note 7**). Exsanguinate the rats by cutting the carotid artery/jugular vein to decrease the chances of false-positive samples resulting from circulating virus. Dissect the animal and collect the relevant organs. Take care not to cross-contaminate organs; harvest them in an order starting from one that is the least likely to contain virus and collect the tumor last. Snap-freeze the organs in liquid N_2; store them at –80°C.
4. Homogenize the samples (*see* **Note 8**).
5. To release virions from the cells, freeze–thaw the suspension three times and sonicate as described in Chapter 18.
6. Clarify the lysate by pelleting the debris in a microcentrifuge at top speed for 1 min.
7. Transfer the supernatant into a fresh tube and discard the pellet.
8. Determine the titer of the virus by a plaque assay (*see* Chapter 18) or a $TCID_{50}$ assay (*see* Chapter 14).

4. Notes

1. LCRT cells were not cloned and appear heterogeneous in cell culture. The fast-growing subpopulation forms clumps of small cells that attach loosely to the surface.
2. A harvest time right after the infection incubation period tells how effectively the nonattached virions were washed away. Early time points (1–4 d p.i.) provide important information about the speed of virus replication. In our experience, wt Ad yield reaches a plateau by 4 d p.i. However, including a later harvest time point (i.e., at 7–8 d) is a good safety measure.
3. Humidification is very important as loss of medium by evaporation will influence the outcome of the experiment.
4. These experiments are best performed in multiwell (we prefer 48-well) dishes so that several Ads can be compared across a range of multiplicities. Arrange the different MOI for a given virus within a column with test and control viruses in parallel columns. Thus, a row will contain wells infected with the same multiplicity for each virus, facilitating comparison.
5. Do not aspirate all of the medium from the wells, because wells are prone to drying out in the laminar flow hood. Also, aspirate medium from only one column at a time.
6. Be careful not to aspirate the cells. When using a vacuum-operated apparatus, do not touch the tip of the aspirating pipet to the bottom of the well.
7. An early time point (e.g., 2 h postinjection) is useful to determine the "leaking out" of the virus from the tumor. Chances to detect virus replication in the tumor and target organs are better from day 1 to day 4 or 5 p.i., while a late time point (e.g., 14 d) could address established infection.
8. For a small number of samples, using the Bio-Pulverizer device is a good choice. This apparatus uses a metallic mortar–pestle assembly chilled in liquid N_2. The

frozen sample is positioned in the mortar and is pulverized by pounding on the pestle with a plastic hammer. The powder then is transferred to a fresh tube. It is important to chill the device completely; otherwise, the sample will thaw and stick to the mortar. For a greater number of samples we would suggest using a bead-beater type homogenizer like the Qiagen TissueLyser, which can process 48 samples at the same time. The caveat of this system is that one cannot process bulkier organs like the liver in one container; these organs need to be split up into multiple tubes. Alternatively, a representative sample of the organ small enough to fit into a tube can be taken. With this device, the samples are put into 1.7-mL screw-cap (with an O-ring) tubes. We use BioStor (United Laboratory Plastics, St. Louis, MO) tubes; these are specially designed to endure freezing in liquid N_2, and they withstand the vigorous shaking in the apparatus. In addition to the sample, a tungsten-carbide ball and 500–1000 µL PBS (depending on the sample size, the total volume should be between 500–1500 µL) is added to the tube. The tube is then placed into the shaker device and shaken back and forth at 30 Hz for 10 min. After homogenization, the ball is removed from the tube; because it is heavier than the organ suspension, it can be "poured" out of the tube.

References

1. Berencsi, K., Uri, A., Valyi-Nagy, T., et al. (1994) Early region 3-replacement adenovirus recombinants are less pathogenic in cotton rats and mice than early region 3-deleted viruses. *Lab. Invest.* **7**, 350–358.
2. Ginsberg, H. S., Lundholm-Beauchamp, U., Horswood, R. L., et al. (1989) Role of early region 3 (E3) in pathogenesis of adenovirus disease. *Proc. Natl. Acad. Sci. USA* **86**, 3823–3827.
3. Pacini, D. L., Dubovi, E. J., and Clyde, W. A., Jr. (1984) A new animal model for human respiratory tract disease due to adenovirus. *J. Infect. Dis.* **150**, 92–97.
4. Prince, G. A., Porter, D. A., Jenson, A. B., Horswold, R. L., Chanock, R. M., and Ginsberg, H. S. (1993) Pathogenesis of adenovirus type 5 pneumonia in cotton rats (sigmodon hispidus). *J. Virol.* **67**, 101–111.
5. Tsai, J. C., Garlinghouse, G., McDonnell, P. J., and Trousdale, M. D. (1992) An experimental animal model of adenovirus-induced ocular disease. The cotton rat. *Arch. Ophthalmol.* **110**, 1167–1170.
6. Toth, K., Spencer, J. F., Tollefson, A. E., et al. (2005) Cotton rat tumor model for the evaluation of oncolytic adenoviruses. *Hum. Gene Ther.* **16**, 139–146.
7. Doronin, K., Toth, K., Kuppuswamy, M., Ward, P., Tollefson, A. E., and Wold, W. S. M. (2000) Tumor-specific, replication-competent adenovirus vectors overexpressing the Adenovirus Death Protein. *J. Virol.* **74**, 6147–6155.
8. Doronin, K., Kuppuswamy, M., Toth, K., et al. (2001) Tissue-specific, tumor-selective, replication-competent adenovirus vector for cancer gene therapy. *J. Virol.* **75**, 3314–3324.
9. Doronin, K., Toth, K., Kuppuswamy, M., Krajcsi, P., Tollefson, A. E., and Wold, W. S. M. (2003) Overexpression of the ADP (E3-11.6K) protein increases cell lysis and spread of adenovirus. *Virology* **305**, 378–387.

10. Toth, K., Djeha, H., Ying, B. L., et al. (2004) An oncolytic adenovirus vector combining enhanced cell-to-cell spreading, mediated by the ADP cytolytic protein, with selective-replication in cancer cells with deregulated Wnt signaling. *Cancer Res.* **64,** 3638–3644.
11. Faith, R. E., Montgomery, C. A., Durfee, W. J., Aguilar-Cordova, E., and Wyde, P. R. (1997) The cotton rat in biomedical research. *Lab. Anim. Sci.* **47,** 337–345.
12. Ward, L. E. (2001) Handling the cotton rat for research. *Lab Animal* **30,** 45–50.
13. Toth, K., Tarakanova, V., Doronin, K., et al. (2003) Radiation increases the activity of oncolytic adenovirus cancer gene therapy vectors that overexpress the ADP (E3-11.6K) protein. *Cancer Gene Ther.* **10,** 193–200.

13

Use of the Syrian Hamster as an Animal Model for Oncolytic Adenovirus Vectors

Maria A. Thomas, Jacqueline F. Spencer, and William S. M. Wold

Summary

Oncolytic adenoviruses (Ads) are promising candidates for cancer therapy. However, current animal models to evaluate these vectors have substantial limitations. Because Ad replication is generally species-specific, oncolytic Ads are usually examined in immunodeficient mice bearing human xenograft tumors. However, this model suffers because the animals are immunodeficient and normal and cancerous mouse tissues are poorly permissive to human Ad replication. We have recently developed a Syrian hamster model that is both immunocompetent and permissive to human Ad replication in normal and cancerous tissues. The Syrian hamster is also permissive for Ad5 replication in the lung, which is the natural site of infection in humans. Human Ads replicate well in vitro in the Syrian hamster cell lines examined and demonstrate significant antitumor efficacy following injection into Syrian hamster tumors in vivo. In this chapter we describe the maintenance of these Syrian hamster cell lines in culture and how to assess oncolytic Ad vector replication in these cells in vitro. We also describe detailed methods for growth of these cell lines as subcutaneous tumors, for intravenous and intratumoral injections in hamsters, and for evaluation of the efficacy, replication, and biodistribution of oncolytic Ad vectors following administration in hamsters. In addition, we describe how to assess replication in normal tissues such as the lungs and give helpful tips on handling, anesthesia, and general care of Syrian hamsters.

Key Words: Syrian hamster; adenovirus; oncolytic; cancer; vector; animal model; tumor; permissive; immunocompetent; replication; in vivo; gene therapy.

1. Introduction

Oncolytic adenoviruses (Ads) are promising candidates for cancer therapy. However, current animal models to evaluate these vectors have substantial limitations. Because Ad replication is generally species-specific (*1*), oncolytic Ads are usually examined in immunodeficient mice bearing human xenograft

From: *Methods in Molecular Medicine, Vol. 130:*
Adenovirus Methods and Protocols, Second Edition, vol. 1:
Adenoviruses Ad Vectors, Quantitation, and Animal Models
Edited by: W. S. M. Wold and A. E. Tollefson © Humana Press Inc., Totowa, NJ

tumors. However, this model suffers because the animals are immunodeficient and normal and cancerous mouse tissues are poorly permissive to human Ad replication *(2–6)*. We have recently developed a Syrian hamster model that is both immunocompetent and permissive to human Ad replication in normal and cancerous tissues *(7)*. The Syrian hamster is also permissive for Ad5 replication in the lung *(5,7,8)*, which is the natural site of infection in humans. Human Ads replicate well in vitro in the Syrian hamster cell lines examined and demonstrate significant antitumor efficacy following injection into Syrian hamster tumors in vivo *(7)*. This chapter describes the maintenance of these Syrian hamster cell lines in culture and how to assess oncolytic Ad vector replication in these cells in vitro. We also describe detailed methods for growth of these cell lines as subcutaneous tumors, for intravenous and intratumoral injections in hamsters, and for evaluation of the efficacy, replication, and biodistribution of oncolytic Ad vectors following administration in hamsters. In addition, we describe how to assess replication in normal tissues, such as the lungs, and give helpful instructions on handling, anesthesia, and general care of Syrian hamsters.

2. Materials

1. Dulbecco's modified Eagle's medium (DMEM) (JRH Biosciences, Lenexa, KS).
2. RPMI-1640 medium (Hyclone, Logan, UT).
3. Fetal bovine serum (FBS) (Hyclone, Logan, UT).
4. HaK cell line (ATCC, Manassas, VA).
5. DDT1 MF-2 cell line (ATCC, Manassas, VA).
6. PC1 cell line (obtained from Dr. Parviz Pour, University of Nebraska Medical Center, Omaha, NE) *(9)*.
7. Syrian hamsters (*Mesocricetus auratus*), 4- to 5-wk-old females (Harlan Sprague Dawley, Indianapolis, IN).
8. Polycarbonate mouse/hamster cages (Allentown Caging, Allentown, NJ).
9. Tekfresh rodent bedding (Harlan Sprague Dawley, Indianapolis, IN).
10. Teklad 2018 rodent food (Harlan Sprague Dawley, Indianapolis, IN).
11. Tuberculin syringe with 25G × 5/8" needle (Becton Dickson, Franklin Lakes, NJ).
12. 45 mg/mL ketamine and 5 mg/mL xylazine in water.
13. Puralube petrolatum ophthalmic ointment (Fougera, Melville, NY).
14. Circulating water blanket (Jorgensen Laboratories, Loveland, CO).
15. Rat intubation pack (Braintree Scientific, Braintree, MA).
16. Sterile polyethylene tubing, PE 50 (Becton Dickinson, Parsippany, NJ).
17. Monoject tuberculin syringe, 1.0 mL (Sherwood Medical, St. Louis, MO).
18. Sterile 23G blunt needle (Braintree Scientific, Braintree, MA).
19. Rodent work stand (Braintree Scientific, Braintree, MA).
20. Micropore surgical tape (3M, St. Paul, MN).
21. Welsh Allyn LI ion otoscope (Braintree Scientific, Braintree, MA).
22. 2% Lidocaine.
23. PowerGen Model 125 handheld homogenizer (Fisher, Pittsburgh, PA).

24. OMNI-Tip Plastic Disposable Rotor Stator Probes (Fisher, Pittsburgh, PA).
25. Syringe filters (such as Acrodisc 13, 0.2 µm, low protein binding, Gelman no. 4454).
26. Monoject tuberculin syringe, 0.5 mL with attached 28G × 1/2" needle (Sherwood Medical, St. Louis, MO).
27. Peanut clippers (Wahl, Sterling, IL) or other hair clippers.
28. Sylvac digital calipers (Fowler, Newton, MA).
29. Insulin syringe, 0.3 mL with attached 29G × 1/2" needle (Terumo, Summerset, NJ).
30. Sterile BioStor vials assembled with screw caps, 1.6 mL (United Laboratory Plastics, St. Louis, MO).
31. TissueLyser homogenizer (Qiagen, Valencia, CA).
32. Tungsten carbide beads, 3-mm (Qiagen, Valencia, CA).

3. Methods

The following methods will be described: the cell culture of Syrian hamster cell lines (**Subheading 3.1.**), assessment of Ad replication in hamster cells (**Subheading 3.2.**), handling and care of hamsters (**Subheading 3.3.**), evaluation of Ad replication in the lungs (**Subheading 3.4.**), intravenous injection of Ad in hamsters (**Subheading 3.5.**), growth of subcutaneous tumors in hamsters (**Subheading 3.6.**), and determination of Ad efficacy, replication, and biodistribution following Ad administration in hamsters (**Subheading 3.7.**).

3.1. Cell Culture of Syrian Hamster Cell Lines

3.1.1. HaK Cells

1. HaK cells can be maintained in DMEM containing 10% (v/v) FBS.
2. When the monolayer nears 100% confluency, subculture the dish. Aspirate the medium and add trypsin to cover the bottom of the dish (*see* **Note 1**).
3. When cells are rounding up and detaching from the dish, add an equal amount of medium and disperse the cells by gently pipetting up and down across the dish.
4. Add the appropriate volume of medium containing cells to a new dish to obtain the desired split ratio (1:10 to 1:20 for HaK). Bring the medium up to a total volume of 10 mL (for a 100-mm diameter dish) by adding DME containing 10% FBS to the new dish. Incubate at 37°C.

3.1.2. DDT1 MF-2 Cells

1. DDT1 MF-2 cells can be maintained in DMEM containing 10% (v/v) FBS (*see* **Note 2**).
2. Subculture DDT1 MF-2 cells at a ratio of 1:10 to 1:20 (*see* **Subheading 3.1.1.**, **steps 2–4**).

3.1.3. PC1 Cells

1. PC1 cells can be maintained in RPMI-1640 medium containing 10% (v/v) FBS (*see* **Note 3**).
2. Subculture PC1 cells at a ratio of 1:3 to 1:6 (*see* **Subheading 3.1.1.**, **steps 2–4**).

3.2. Assessment of Ad Replication in Hamster Cells In Vitro

Ad replication can be assessed by performing a single-step growth curve in which a high multiplicity of infection is used such that a single round of infection will occur. The virus yields obtained in different cell lines or with different viruses can be compared by determining the plaque-forming units (PFU) produced per cell (PFU/cell).

3.2.1. Infection

1. Plate cells (HaK, PC1, or DDT1 MF-2) onto 35-mm dishes such that monolayers will be 50–70% confluent in 1–2 d. The number of dishes needed for each cell line is one dish per time point per virus, plus one dish per cell line for cell counting.
2. When monolayers are 50–70% confluent, remove the medium from one dish per cell line and trypsinize the monolayer. Determine the total cell number per dish by counting the cells on a hemacytometer. Multiply the number of cells per dish by the multiplicity of infection (i.e., 100 PFU/cell) to determine the PFU needed per dish.
3. Prepare a dilution of the virus stock to be examined in serum-free medium such that the inoculum will be between 10 and 100 μL per dish.
4. Remove the medium from the dishes and add 0.5 mL of serum-free medium. Add the virus directly into the serum-free medium. Incubate the dishes at 37°C. Gently rotate the dishes every 15 min.
5. At 1 h postinfection (p.i.), remove the medium from the dishes. Wash the monolayers three times with serum-free medium. Add 2 mL medium containing 5% FBS and return the dishes to the incubator.

3.2.2. Virus Recovery

1. At each time point, collect the cells and medium from one dish and freeze at –80°C.
2. Once all time points have been collected, process all the samples as follows. Freeze (at –80°C) and thaw the samples a total of three times each. Bring all samples up to the same total volume (i.e., 2 mL).
3. Following the final thaw, sonicate the samples three times (3 min each time) in a cup sonicator cooled with recirculating 4°C water, keeping the samples on ice when not being sonicated.
4. Clarify the lysates by centrifugation (3 min at 16,000*g*) prior to titering the virus yield. Save both the supernatant and the pellet.

3.2.3. Determination of the Virus Burst Size

1. Following centrifugation of the crude lysates prepared above, titer the virus yield by plaque assay of the supernatant. This will give the yield in PFU per mL of sample.

2. Multiply the titer obtained by plaque assay (PFU/mL) by the total mL of sample to determine the total PFU. Calculate the burst size (PFU/cell) for each sample by dividing the total PFU by the total number of cells per dish at the time of infection. The burst sizes may then be compared for different cell lines and/or different viruses.

3.3. General Care and Handling of Syrian Hamsters

3.3.1. Housing and General Care of Hamsters

1. Hamsters should be housed two to three animals per cage (*see* **Note 4**) in polycarbonate mouse/hamster cages with rodent bedding placed in the bottom of the cage for nesting. Use a stainless steel wire bar lid that snaps down tightly to prevent hamsters from escaping.
2. Use relatively slow and deliberate movements when handling hamsters. Hamsters are relatively easy to handle, but may be momentarily aggressive when suddenly startled or awakened by a handler.
3. The animals should be fed a standard rodent diet containing at least 18% protein and should have free access to water. Hamsters seem to tolerate shipping well, but food and water consumption should be monitored during acclimation in case of shipping stress. An acclimation period of at least 3 d is recommended prior to any interventions.

3.3.2. Physical Restraint

1. For husbandry purposes, the animals can be picked up by grasping the scruff of the hamster's neck between your thumb and index finger.
2. For procedures that require immobilization, the animals may be physically restrained by scruffing the skin over the shoulders and back tightly, while pushing the animal down with an appropriate amount of force. Hamsters have very "loose" skin, and the handler must scruff as much skin as possible to ensure good restraint.

3.3.3. Anesthetic Restraint

Anesthesia may be necessary for procedures that require the animal to be immobilized for an extended period of time such as for intravenous injection or endotracheal instillation of virus.

1. First, check and empty the hamster's mouth and cheek pouches prior to administering anesthesia.
2. Draw up the anesthetic in a tuberculin syringe with a 25G × 5/8" needle. An anesthetic cocktail containing 45 mg/mL ketamine and 5 mg/mL xylazine in water can be administered at a dose of 0.2 mL/100 g body weight for intravenous injection or at 0.4 mL/100 g body weight for endotracheal instillation (for male hamsters, *see* **Note 5**).

3. The hamster should be scruff-restrained, as previously described, and placed in dorsal recumbency with the head pointing downward.
4. Insert the needle into the lower right quadrant of the abdomen at an approx 45° angle. Before depressing the plunger, pull back slightly on the plunger to ensure correct needle placement. If blood, feces, or urine is aspirated, withdraw the needle, discard the syringe, and repeat the procedure using a new syringe of anesthetic cocktail.
5. Once proper placement has been determined, administer the anesthetic by intraperitoneal injection.
6. Place petrolatum ophthalmic ointment in the animal's eyes to prevent corneal dryness and abrasions.
7. Place the animal in a cage that has been placed on a prewarmed circulating water blanket until it loses consciousness. An anesthetized animal's airway can become blocked if it aspirates bedding; therefore, cages without bedding are used for both anesthesia induction and recovery.
8. Prior to performing a procedure, assess the depth of anesthesia by pinching between the toes for pedal reflex. The animal should not pull its foot away or have any contracting muscular response to this stimulus. If this reflex is absent, the animal is at an appropriate anesthetic depth for most noninvasive procedures. If this reflex is present, more time may be needed for the anesthesia to take effect. If the animal is not fully anesthetized after 7–10 min, it can be redosed with anesthesia using one-quarter to one-half the previous dose and subsequently monitored until it is at an appropriate anesthetic depth.
9. Once the appropriate anesthetic depth is attained, perform the desired procedure (such as intravenous injection or endotracheal instillation of virus).
10. During anesthetic recovery, place the hamster in a cage without bedding that has been placed on a prewarmed circulating water blanket.
11. Once the animal is mobile and has a righting reflex, it can be returned to its original cage.

3.4. Evaluation of Ad Replication in Hamster Lungs

The natural site of Ad5 replication in humans is the lung, where it causes a mild respiratory illness in young children. In order to evaluate the replication of an oncolytic Ad vector in the lungs and thus the cancer-specificity of your vector, inoculate the hamster by endotracheal instillation of the vector, harvest the lungs, and determine the total virus yield from the lungs.

3.4.1. Endotracheal Instillation of Virus

1. Prior to performing this procedure, it may be helpful to view the instructional video that is provided in the rat intubation pack (**Heading 2.**, **item 15**) and practice the procedure with saline.
2. For instillation of virus, polyethylene tubing is attached to a 1-mL tuberculin syringe by a 23-gage blunt needle. Prior to instillation, you will need to deter-

mine the appropriate length for this tubing. Do this by measuring the distance from the nose to the thoracic inlet with a piece of tubing. Align the end of the tubing at the level of the thoracic inlet and mark the tubing where it reaches the nose. With a razor blade, cut the tubing approx 1" beyond this mark to allow for the needle to be inserted (*see* **Note 6**).

3. Prepare the virus at the desired concentration in phosphate-buffered saline (PBS). We have administered 1×10^7 PFU in a volume of 0.1 mL. Prepare enough for the number of animals in the study, plus approx 50% extra.

4. Anesthetize the hamster (*see* **Subheading 3.3.3.**).

5. With the blunt needle and syringe described above, first draw up 0.2–0.3 mL of air, which will be used to flush out the full volume of virus through the length of the tubing following instillation. Next, draw up 0.1 mL of your virus preparation. Then, attach the tubing to the blunt needle.

6. Once the animal is at an appropriate anesthetic depth, place the animal in dorsal recumbency on the rodent work stand with the platform parallel to the table. Use the tooth loop, side arms, and Micropore surgical tape (if necessary) to secure the animal in place.

7. Fully elevate the platform and position it so that the animal's head is toward the handler, and the ventrum of the animal is facing away from the handler.

8. With your right (or dominant hand), place a cotton tip applicator on the dorsal surface of the tongue and roll it up and out. Keep the tongue extended during intubation. Then, with your left (or nondominant) hand, gently insert the otoscope into the mouth (*see* **Note 7**). This hand will keep the otoscope in position for the remainder of the procedure while your other hand performs other functions (lidocaine application, placement of tubing, and instillation of virus).

9. Once the otoscope is positioned inside of the mouth, the vocal chords should be in view. Cricoid pressure may be applied to help facilitate visualization of the chords. If excess fluid is present in front of the chords, a cotton-tipped applicator can be used to gently remove fluids under direct visualization.

10. One to two drops of lidocaine should then be applied to the vocal chords with your right (or dominant) hand, using the lidocaine applicator supplied in the intubation pack. Use a cotton-tipped applicator to remove excess lidocaine prior to instillation, if necessary.

11. While visualizing the vocal chords, gently insert the tubing between the vocal folds with your right (or dominant) hand. Time the insertion of the tube with the opening of the vocal folds during the animal's respiration. Do not force the tube between the chords. Once fully inserted, the mark on the tubing should rest at the corner of the hamster's mouth.

12. Inject the virus (suspended in no more than 0.1 mL of vehicle) slowly through the tubing. Follow the virus instillation by the air drawn into the syringe to clear the tubing of fluid.

13. After injection, the tube can be withdrawn from the mouth, followed by careful withdrawal of the otoscope. Intubation, injection, and withdrawal of the tube should all be visualized through the otoscope to ensure correct placement and confirm dose administration.

14. Recover the animals (*see* **Subheading 3.3.3.**) with the following modifications. The animals should be recovered in an oxygen-filled chamber. Place each animal in sternal recumbency with the head elevated (use a 1/2–1" roll of gauze under the neck of each animal or elevate the appropriate side of the entire cage) until fully recovered. Monitor the animals and obtain the assistance of a veterinarian if complications arise.

3.4.2. Lung Harvest and Virus Yield Determination

1. At various timepoints postinstillation of virus, animals are sacrificed and lungs are harvested. To minimize blood contamination of the lungs, exsanguinate or perfuse the vascular system prior to harvesting the lungs.
2. Dissect the lungs from each animal, and place in a 100-mm dish on ice. In a hood, trim away any excess tissue. Note any gross pathology observations.
3. Determine the weight of a new 100-mm dish, and mince the lungs with a disposable scalpel in this dish until no large chunks remain. Weigh the minced lungs in this dish, and calculate the net lung weight.
4. Place each lung homogenate into a 6-mL (12 × 75 mm) snap-cap tube. Add 0.5 mL PBS to the dish to recover any small chunks left on the dish. Add this PBS to the rest of the sample in the tube.
5. Place the tubes in a dry ice–ethanol bath to freeze. Freeze samples (at –80°C) until tissues from all time points have been collected.
6. Homogenize each sample with a handheld homogenizer (such as the PowerGen Model 125) equipped with a disposable rotor stator probe. Use a different probe in each sample (*see* **Note 8**).
7. Freeze (at –80°C) and thaw the samples three times total.
8. Sonicate the samples three times (3 min each time), keeping the samples on ice when not being sonicated.
9. Bring each sample up to the same total volume (i.e., 1.5 mL) with PBS.
10. Clarify the homogenates by centrifugation (15 min at 2060g) prior to titer determination.
11. Determine the virus yield for each sample by plaque assay of the supernatant (*see* **Note 9**). Multiply the final titer (PFU/mL) by the total sample volume to determine the total PFU recovered from the lungs per animal at each time point.

3.5. Intravenous Injection of Ad in Syrian Hamsters

Intravenous injection of virus may be desired to determine the maximum tolerated dose or the efficacy of an oncolytic adenovirus vector when administered systemically. In our experience, the jugular vein is the preferred route for intravenous injection in the hamster. The vessel is easily accessible and using this route of administration poses little risk of accidental death or injury to the animal.

1. First dilute the oncolytic vector to the desired concentration(s) in PBS for the evaluation of toxicity or efficacy.

2. Anesthetize the hamster (*see* **Subheading 3.3.3.**). Draw up the virus to be administered in a 0.5-mL tuberculin syringe with an attached 28G × 1/2" needle.
3. When the animal is almost fully anesthetized, shave the ventral neck area with hair clippers from below the clavicles to midway up the neck.
4. When the hamster is fully anesthetized, swab the injection site with 70% ethanol.
5. Place the animal on a rodent work stand in dorsal recumbency. Place a small roll of gauze (1/2–3/4" in diameter) under the base of the hamster's neck so that the head is slightly hyperextended. Loop the upper front teeth with the tooth loop provided with the stand, and attach the other end of the tooth loop to the stand. To better facilitate access to the vessel, use surgical tape to affix the front legs to the side rails of the stand.
6. Determine the location of the jugular vein approximately at the midpoint of the clavicle. If the clavicle and jugular vein cannot be visualized, it may be necessary to palpate the clavicle to determine the midpoint. The salivary glands may be gently manipulated cranially or laterally if they are obstructing the injection site.
7. Position the syringe at a 45° angle, immediately cranial to the clavicle. Enter the skin with the needle bevel-up, and pull back on the plunger gently as the needle is advanced. As the needle enters the vessel, blood will flow into the barrel of the syringe (*see* **Note 10**). Once blood is flowing into the syringe barrel, the syringe contents may be slowly injected.
8. While injecting, keep the syringe position stable. Any repositioning of the needle may result in perivascular injection. If the syringe plunger becomes harder to advance or a raised area becomes visible at the injection site, the syringe may need to be repositioned and the patency of the needle checked.
9. After injecting the contents of the syringe, remove the needle and hold off the vessel using mild pressure for approx 5–10 s.
10. Gently remove the animal from the stand and remove the tooth loop. Recover the animal from anesthesia (*see* **Subheading 3.3.3.**).
11. Depending on the purpose of the study, monitor the animals for toxicity (if performing a maximum tolerated dose study) or assess the antitumor efficacy of the oncolytic adenovirus vector (if performing an intravenous efficacy study).

3.6. Growth of Subcutaneous Tumors in the Syrian Hamster

3.6.1. Preparation of Tumor Cells

1. Expand the cells in multiple 150-mm diameter dishes. The number of dishes needed will depend on the number of tumors desired (*see* **Note 11**).
2. When dishes are confluent, remove the medium and trypsinize the cells (*see* **Subheading 3.1.1.**). For each dish, add 5 mL of trypsin and, when the cells are detached, add 5 mL of medium. Combine the cells from all of the dishes (10 mL per dish) into one tube (i.e., 250-mL tube, if 20 dishes).
3. Determine the cell concentration by performing a cell count on a hemacytometer. Multiply the cell concentration (cells/mL) by the total volume (i.e., 200 mL, if 20 dishes) to determine the total number of cells harvested.

4. Pellet the cells by centrifugation (10 min at 228*g*). Pour off most of the medium, and then carefully aspirate the remaining medium. Rinse the cells by resuspending them in 200 mL serum-free medium.

5. Pellet the cells again as in **step 4**. Remove the medium, and resuspend the cell pellet to yield a total volume such that the desired cell concentration is attained (*see* **Note 12**).

3.6.2. Subcutaneous Injection of Tumor Cells

1. Prior to cell injection, the area where cells will be injected should be shaved with hair clippers, being careful not to nick the skin. The animals can be shaved the day before cell injection if desired. Typically the hindflanks are used, with tumors grown on one or both flanks (*see* **Note 13**).

2. For cell injection, have one person restrain the hamster while the other person injects the cells. The restrainer should place the palm of his/her hand over the head of the animal while wrapping his/her fingers and thumb around the thorax just behind the shoulder blades. The restrainer should be sure not to impede the animal's ability to breath by restraining too tightly. If correctly restrained, the animal's front legs will be pushed cranially, the top of the head will rest in the restrainer's palm, and the lower half of the animal's body will be easily accessible for injection. It may be necessary for the restrainer to hold the back legs of the animal with his/her other hand. A table or cage may be used to support the restrainer's hands and the animal. Proper restraint will limit squirming and keep the cells from being injected in multiple locations.

3. Once the animal is restrained, the second person can inject the cells into the hind flank. Draw up the volume of cells to be injected (typically 100–200 μL) in a tuberculin syringe with a 22G × 1" needle. The person injecting the cells should gently pinch some skin with his/her left (or nondominant) hand to make a tent for placement of the needle.

4. With his/her right (or dominant) hand, the person injecting the cells should carefully insert the needle, bevel-side-up, into the center of the tented skin (*see* **Note 14**). Slowly inject the cells so that they are deposited in one area. When the entire volume has been injected, slowly withdraw the needle and discard in a sharps container.

5. Temporarily place the hamster that has been injected with cells in an area separate from its cagemates, i.e., an extra cage or above the wirebar, to keep track of which animals have already been injected with cells.

6. Monitor the development of tumors and shave the animals when necessary to view and palpate the area. Different cell lines form tumors at different rates (*see* **Note 15**).

7. Because the tumors are very small, they are somewhat difficult to palpate because hamsters have thick abundant skin. If intratumoral injection is used, wait until the tumors can be easily found, measured, and injected before beginning the treatment.

3.7. Determination of Efficacy, Replication, and Biodistribution Following Ad Administration

Because the hamster is a replication-permissive immunocompetent model, the interactions between an oncolytic Ad vector, the tumor, and the host can be examined. Following intratumoral injection of vector, the efficacy of oncolytic Ads can be compared. Also, the intratumoral replication of oncolytic Ads as well as the replication and biodistribution of the vector in normal tissues can be evaluated in this model. These events can be studied following intravenous administration as well (for intravenous injection, *see* **Subheading 3.5.**).

3.7.1. Efficacy With Intratumoral Ad Injection

1. There may be significant variation in tumor size among animals. It is best to measure the tumors (*see* **Note 16**) prior to the initiation of treatment and randomize the animals into treatment groups such that the mean tumor volume of each group is similar.
2. Decide on the desired dose and injection volume (*see* **Note 17**) and prepare your vector accordingly by dilution in vehicle (e.g., PBS) to yield the desired concentration. Prepare the amount of virus needed for all of the tumors to be injected that day, plus approx 20% extra. A vehicle-only injection is used as a control to compare the effect of oncolytic Ad injection on tumor growth.
3. The area surrounding the tumor should be shaved prior to virus injection.
4. Intratumoral virus injection in hamsters can be performed by one person. First, draw up the amount of virus to be injected in a 0.3-mL insulin syringe with an attached 29G × 1/2" needle. With your left (or nondominant) hand, scruff the skin of the hamster by taking hold of as much of the skin around the animal's cheeks, neck, dorsum, and opposite flank as possible to reduce squirming and the possibility of being bitten or of injecting yourself.
5. Once the animal is restrained, use your other hand to inject the vector directly into the tumor. Dispense the volume slowly and throughout the tumor by slightly withdrawing the needle (do not pull out of the tumor entirely) and making several needle tracks throughout the width and depth of the tumor (*see* **Note 18**).
6. When the entire volume has been injected, slowly remove the needle from the tumor and discard in a sharps container. Then place the animal that has been injected in an area separate from its cagemates, e.g., on top of the wirebar, until all animals in a cage have been injected.
7. Continue the dosing schedule on subsequent days as desired. Measure the tumors twice a week with digital calipers to monitor tumor growth and evaluate the effect of vector treatment. Periodically it will be necessary to shave the area around the tumor prior to tumor measurement to ensure accuracy.

3.7.2. Replication and Biodistribution Following Intratumoral Ad Injection

1. To observe vector replication and biodistribution, a single vector injection (intratumoral or intravenous) should be administered, as described previously (*see* **Subheading 3.7.1.** or **3.5.**, respectively). The use of a replication-defective vector is a valuable control for vector clearance.

2. At various time points postinjection of vector, sacrifice the number of animals to be analyzed (at least three per group). Blood may be collected by cardiac puncture if desired. Exsanguinate or perfuse the animals to reduce blood contamination of the collected organs.

3. Use separate instrument sets to dissect the skin away, to open the abdominal and thoracic cavities and exsanguinate, and to dissect the organs. Collect the organs to be analyzed in the order of (probable) increasing virus load (i.e., lungs, then liver, and tumor last). Place the organs in dishes on ice and disinfect the instrument sets between animals.

4. In a hood, trim any excess tissue away from the harvested organs and chop the larger organs such that they will fit into a 1.6-mL screw-cap tube. Tare the weight of one screw-cap tube (with cap), transfer each organ into a tube, and determine the net weight of each organ (*see* **Note 19**).

5. Freeze samples in a dry ice–ethanol bath and store at −80°C until samples from all time points have been collected.

6. Homogenize the organs in PBS with a Qiagen TissueLyser bead mixer, using one tungsten carbide bead per tube. After homogenization, remove the bead from each tube and recombine multiple tubes of the same organ if necessary. If the sample has a thick consistency, the bead can be removed by pouring the bead out.

7. Freeze (at −80°C) and thaw the samples three times each. Sonicate the samples two times (4 min each). Add PBS to bring all samples of a given organ up to the same final volume to ease final yield calculations.

8. Prior to titering, clarify each sample by centrifugation (blood for 15 min at 16,000g; livers, lungs, tumors for 15 min at 2060g).

9. Titer the supernatant of each sample by TCID$_{50}$ assay (*see* **Note 20**) on 293 cells (if a replication-defective control was included).

10. Multiply the titer obtained from the TCID$_{50}$ assay by the total volume of sample to yield the total infectious units per organ.

4. Notes

1. HaK cells adhere very strongly to tissue culture dishes, and thus trypsinization is quite slow. To optimize trypsinization, subculture HaK cells before they reach 100% confluency, rinse the monolayer one to two times with PBS or trypsin, and incubate the plate at 37°C during trypsinization.

2. DDT1 MF-2 cells grow both as adherent cells and in suspension and are easily trypsinized.

3. PC1 cells grow as well-adherent islands of cells. Follow the suggestions for efficient trypsinization of HaK cells (*see* **Note 1**).

4. In our experience, hamsters will fight to gain territory in the cage for sleeping. A maximum of three hamsters per cage is thus recommended.

5. Male hamsters tend to be affected more by the anesthetic and may require a lesser dose. In addition, when using male hamsters, the handler should be careful not to puncture the right testis when inserting the needle, as the testis may have been pushed up into the abdominal cavity during restraint.

6. Proper tubing length will ensure instillation within the trachea and will prevent the tube from being inserted past the bronchial bifurcation, which would result in asymmetrical instillation of virus to the lungs.

7. The handler must be careful not to move the otoscope from side to side or force the otoscope into the throat. Rough handling with the otoscope can lead to soft tissue lacerations and swelling. If either of these conditions occurs, the procedure on this animal should not continue, and the animal should be closely monitored for signs of distress.

8. The OMNI-Tip probes can be reused if, after use, they are rinsed in water, washed briefly in 10% bleach, rinsed again in tap water, and finally rinsed with distilled water.

9. Because the lung tissue cannot be considered to be "sterile," the solution for plaque assay is passed through a 0.2-μm filter after the first dilution. Adenovirus virions are approx 0.1 μm and would be expected to pass through the filter.

10. If the needle has been advanced more than three-quarters of the way and no blood has entered the syringe, the needle has either missed or advanced through the vessel. Keep slight backpressure on the plunger and slowly back the needle out of the animal. The needle may enter the vessel during withdrawal of the syringe.

11. At confluency, one 150-mm dish of cells will yield approx 2×10^7 to 1×10^8 cells.

12. Cells are typically injected in a volume of 100–200 μL per hind flank. To produce tumors, inject the following cell numbers: inject 2×10^7 HaK cells; inject 1×10^7 DDT1 MF-2 cells; inject 1×10^6 PC1 cells.

13. If one tumor per animal is used, place the tumor on the right side of the animal (if you are right-handed) to ease intratumoral injection of the tumor. Keep in mind that if two tumors are grown per animal, the animal may need to be sacrificed when one tumor becomes too large, regardless of the size of the contralateral tumor.

14. The needle should advance easily under the skin, and the correct needle placement should be verified by slightly pulling back on the plunger to ensure the needle is not in a blood vessel. The contents of the syringe should evacuate with minimal pressure exerted on the plunger. If the needle is hard to advance or the syringe contents are difficult to inject, the needle may be incorrectly placed either intradermally or intramuscularly. After the injection, a small raised area may be present at the site.

15. HaK tumors grow steadily and consistently among animals but somewhat slowly. Tumors that are of measurable and injectable size will take 3–5 wk to form. HaK cells, when injected subcutaneously, may also occasionally form intradermal tumors at the site of injection. HaK tumors frequently metastasize to the lungs, kidneys, and renal lymph nodes. At approx 2 mo after injection of HaK cells,

pulmonary metastasis may affect efficient lung function, so be observant for animals that appear dyspneic. These animals should be sacrificed because of excessive pulmonary metastasis. DDT1 MF-2 tumors, on the other hand, grow rapidly. Injectable tumors will be formed in less than 1 wk. PC1 tumors grow at a variable rate, forming injectable tumors between 2 and 4 wk in most animals.

16. The volume (V) of a subcutaneous tumor can be estimated by measuring the length (L) and the width (W) of the tumor with digital calipers and using the following formula (which approximates $4/3 \, \pi \, r^3$): $V = 0.524 \times L \times W^2$.

17. Typically the injection volume is either 100 or 200 µL, depending on the size of the tumor. If most tumors are under 200 µL, use an injection volume of 100 µL; if most tumors are greater than 200 µL, use an injection volume of 200 µL.

18. Even if the animal squirms while you are injecting the tumor, try not to withdraw the needle entirely unless absolutely necessary. If the needle is withdrawn before the entire dose is administered, keep this animal in a separate location and inject the remainder of the dose after waiting at least 5 min. If a second injection is immediately attempted, the vector will usually leak out of the original site.

19. Larger organs such as the liver will either need to be divided into multiple tubes for homogenization and recombined thereafter (titer can thus be reported per organ or per g of organ); alternatively a portion of the organ (e.g., 1 g) could be processed (titer could thus be reported per g of organ only). Assaying only a portion of the organ assumes that vector distribution was uniform throughout the organ.

20. Perform the TCID$_{50}$ assay for each sample as follows. Plate 293 cells onto 96-well plates (one plate per sample or control vector stock; called the "infection plates") in 100 µL of DMEM containing 10% FBS such that cells will be approx 30% confluent the following day. On the day of infection, add serum-free DMEM to the plates (add 100 µL to Row A [mock infection] and 50 µL to Rows B to H). Perform serial dilutions (20 µL into 180 µL serum-free DMEM) of each sample in a parallel 96-well plate (the "dilution plate"), pipetting up and down and changing tips with each dilution. The dilutions can be performed with a multichannel pipette in Rows C through H (wells 1–6) of the dilution plate, which will result in a 10^{-1} to 10^{-6} dilution series. On the infection plate, pipet 50 µL of undiluted sample into wells 1–12 of Row B (Row A is the mock infection row). Pipet 50 µL of each dilution from the dilution plate into the corresponding row of the infection plate (Rows C to H) with a multichannel pipet. Some samples are toxic to the monolayer (undiluted lungs and tumors, and undiluted and 10^{-1} dilutions of liver samples) and must be removed at 1 h p.i. For these samples only, gently remove the infecting inoculum and add 200 µL of DMEM containing 5% FBS. At 7 d p.i., add approx 150 µL DMEM containing 5% FBS to the plates (but do not remove any medium). At 14 d p.i., score each well (+ or −) for cytopathic effect by microscopic examination. Calculate the titer (TCID$_{50}$ infectious units per mL) by the Reed–Muench method *(10)*.

References

1. Horwitz, M. S. (2001) Adenoviruses, in *Fields Virology* (Knipe, D. M., Howley, P. M., Griffin, D. E., et al., eds), Lippincott Williams & Wilkins, Philadelphia, pp. 2301–2326.
2. Bernt, K. M., Ni, S., Tieu, A. T., and Lieber, A. (2005) Assessment of a combined, adenovirus-mediated oncolytic and immunostimulatory tumor therapy. *Cancer Res.* **65,** 4343–4352.
3. Duncan, S. J., Gordon, F. C., Gregory, D. W., et al. (1978) Infection of mouse liver by human adenovirus type 5. *J. Gen. Virol.* **40,** 45–61.
4. Ginsberg, H. S., Moldawer, L. L., Sehgal, P. B., et al. (1991) A mouse model for investigating the molecular pathogenesis of adenovirus pneumonia. *Proc. Natl. Acad. Sci. USA* **88,** 1651–1655.
5. Hjorth, R. N., Bonde, G. M., Pierzchala, W. A., et al. (1988) A new hamster model for adenoviral vaccination. *Arch. Virol.* **100,** 279–283.
6. Liu, T. C., Hallden, G., Wang, Y., et al. (2004) An E1B-19 kDa gene deletion mutant adenovirus demonstrates tumor necrosis factor-enhanced cancer selectivity and enhanced oncolytic potency. *Mol. Ther.* **9,** 786–803.
7. Thomas, M. A., Spencer, J. F., LaRegina, M. C., et al. (2006) Syrian hamster as a permissive immunocompetent animal model for the study of oncolytic adenovirus vectors. *Cancer Res.* **66,** 1270–1276.
8. Morin, J. E., Lubeck, M. D., Barton, J. E., Conley, A. J., Davis, A. R., and Hung, P. P. (1987) Recombinant adenovirus induces antibody response to hepatitis B virus surface antigen in hamsters. *Proc. Natl. Acad. Sci. USA* **84,** 4626–4630.
9. Egami, H., Tomioka, T., Tempero, M., Kay, D., and Pour, P. M. (1991) Development of intrapancreatic transplantable model of pancreatic duct adenocarcinoma in Syrian golden hamsters. *Am. J. Pathol.* **138,** 557–561.
10. Condit, R. C. (2001) Principles of virology, in *Fields Virology* (Knipe, D. M., Howley, P. M., Griffin, D. E., et al., eds.), Lippincott Williams & Wilkins, Philadelphia, pp. 19–51.

14

A Real-Time PCR Method to Rapidly Titer Adenovirus Stocks

Maria A. Thomas, Drew L. Lichtenstein, Peter Krajcsi, and William S. M. Wold

Summary

A critical step in working with adenovirus (Ad) and its vectors is the accurate, reproducible, sensitive, and rapid measurement of the amount of virus present in a stock. Titration methods fall into one of two categories: determination of either the infectious or the particle (infectious plus noninfectious) titer. Determining the infectious titer of a virus stock by plaque assay has important limitations, including cell line-, researcher-, and laboratory-dependent variation in titer, and the length of time required to perform the assay (2–4 wk). A major drawback of particle titration methods is the lack of consistent correlation between the resultant titer and the infectious titer. To overcome these problems, a rapid, sensitive, and reproducible real-time polymerase chain reaction (PCR) assay was developed that detects encapsidated full-length genomes. Importantly, there is a linear correlation between the titer determined by the real-time PCR assay and the infectious titer determined by a plaque assay. This chapter provides step-by-step guidance for preparing viral DNA, conducting the real-time PCR assay, and using the resultant data to calculate a viral titer.

Key Words: Adenovirus; real-time PCR; titration; SYBR Green; titer; vector.

1. Introduction

A critical step in working with adenovirus (Ad) and its vectors is the accurate, reproducible, sensitive, and rapid measurement of the amount of virus present in a stock. Titration methods fall into one of two categories: determination of either the infectious or the particle (infectious plus noninfectious) titer. Although the plaque assay is a well-established technique for determining the infectious titer of a virus stock, this method has important limitations, including cell line-, researcher-, and laboratory-dependent variation in titer, the long

From: *Methods in Molecular Medicine, Vol. 130:*
Adenovirus Methods and Protocols, Second Edition, vol. 1:
Adenoviruses Ad Vectors, Quantitation, and Animal Models
Edited by: W. S. M. Wold and A. E. Tollefson © Humana Press Inc., Totowa, NJ

length of time required for performing the assay (2–4 wk), and difficulty in detecting the small plaques formed by viruses or vectors that do not express adenovirus death protein (ADP), which is required for the efficient release and thus spread of virions at the culmination of the virus replication cycle *(1)*. A common method for quantifying the particle titer of a stock utilizes the absorbance at 260 nm, which primarily measures viral DNA *(2)*. However, this method most likely measures nonencapsidated DNA as well as incomplete genomes and, therefore, does not accurately reflect the infectious titer. A major drawback of particle titration methods is the lack of consistent correlation between the resultant titer and the infectious titer. With the advent of real-time polymerase chain reaction (PCR), rapid, quantitative titering methods that rely on detection of specific Ad DNA sequences have been reported *(3–5)*. However, because of the method for preparing the DNA and the location of the primers used for amplification, these assays may also detect nonencapsidated DNA and partial genomes. To overcome the problems associated with current titration methodologies, a real-time PCR assay has been developed incorporating a DNase treatment step (to avoid detection of nonencapsidated DNA) as well as primers which are located near the far right end of the genome (to detect only complete genomes). Inclusion of these modifications yields a real-time PCR assay in which the titer of a variety of crude and purified virus stocks is determined in a rapid, sensitive, and reproducible manner. Importantly, there is a linear correlation between the titer determined by the real-time PCR assay and the infectious titer determined by a plaque assay. This chapter provides step-by-step guidance for preparing viral DNA, conducting the real-time PCR assay, and using the resultant data to calculate a viral titer.

2. Materials

1. 10X DNase digestion buffer: 500 mM Tris-HCl, pH 7.6, 100 mM MgCl$_2$.
2. RQ1 RNase-free DNase (Promega, Madison, WI).
3. 10X Proteinase K digestion buffer: 100 mM Tris-HCl, pH 8.0, 100 mM ethylene diamine tetraacetic acid (EDTA), 2.5% (w/v) sodium dodecyl sulfate (SDS).
4. Proteinase K, PCR grade (Roche Applied Sciences, Indianapolis, IN).
5. Wizard DNA Clean-up System (Promega, Madison, WI).
6. 80% (v/v) Ethanol in RNase- and DNase-free water.
7. Luer-Lok syringes, 3-mL.
8. 2X SYBR® Green PCR Master Mix (Applied Biosystems, Foster City, CA).
9. The forward PCR primer (5'-CAGCGTAGCCCCGATGTAA-3') corresponds to base pairs 34,974–34,955 of the Ad serotype 5 (Ad5) reference material (GenBank accession number AY339865). The reverse primer (5'-TTTTTGAG CAGCACCTTGCA-3') corresponds to base pairs 34,955– 34,974 of the Ad5 reference material. Both primers were synthesized by Invitrogen (Carlsbad, CA). Prepare a 1.25 µM working stock of each primer by dilution with RNase- and DNase-free water.

10. Serial 10-fold dilutions (from 1×10^9 to 1×10^2 copies per 7 µL) of the plasmid pJW228, which contains the right-hand end of the Ad5 genome, were prepared for generating a standard curve for real-time PCR.
11. ABI Prism 7700 Sequence Detection System (Applied Biosystems, Foster City, CA) (*see* **Note 1**).
12. MicroAmp optical 96-well reaction plates or MicroAmp optical tubes, 8-tube strips (Applied Biosystems, Foster City, CA).
13. MicroAmp optical adhesive covers or MicroAmp optical caps, 8-cap strips (Applied Biosystems, Foster City, CA).
14. ABI Prism cap installing tool (Applied Biosystems, Foster City, CA).
15. ELIMINase decontaminant solution (Fisher Scientific, Pittsburgh, PA).

3. Methods

We have tested our real-time PCR titration assay with many different virus and vector stocks prepared by different procedures, by different personnel in the same laboratory, and by different laboratories. In addition, the stocks were stored in different buffers and for various periods of time, from weeks to nearly 20 yr. Furthermore, the stocks possessed a variety of different point mutations, insertions, and deletions. In theory, the only viruses that cannot be used in this assay are those that contain a disruption of either primer binding site. The types of stocks assayed included purified virus and crude cell lysate stocks. The buffer components included CsCl, glycerol, and tissue culture medium containing fetal bovine serum. Because PCR-based assays are exquisitely sensitive, we recommend that extreme care be taken when preparing the viral DNA for the assay and in setting up the assay so as to avoid cross-contamination and thus incorrect results (*see* **Note 2**).

3.1. DNase Digestion

Prepare a master mix for the DNase digestion as listed below. The volumes shown are for a single reaction. To simultaneously digest DNA from more than one stock, multiply the following volumes by the number of samples to be digested plus 10% extra (e.g., if there are five samples, then multiply by 5.5): 399 µL sterile RNase- and DNase-free water; 45 µL 10X DNase digestion buffer; 1 µL 0.5U/µL RQ1 DNase (*see* **Note 3**).

1. Dispense 445 µL of the DNase digestion master mix into as many microcentrifuge tubes as there are samples to be purified.
2. Add 5 µL of virus stock to each tube.
3. Incubate the samples at 37°C for 60 min (*see* **Notes 4** and **5**).
4. To inactivate the DNase, incubate the samples at 75°C for 30 min.

3.2. Proteinase K Digestion

1. Prepare a master mix for the Proteinase K digestion as listed below. The volumes shown below are for a single reaction. To simultaneously digest more than one

sample, multiply the following volumes by the number of samples to be digested plus 10% extra: 50 µL 10X Proteinase K digestion buffer; 16 µL Proteinase K (15.6 mg/mL).
2. Add 66 µL of the Proteinase K digestion master mix to each tube.
3. Incubate the samples at 37°C for 60 min (*see* **Note 6**).

3.3. Viral DNA Purification

1. Purify each viral DNA sample with the Wizard DNA Clean-up System according to the manufacturer's recommended protocol, with the exception that 80% ethanol should be substituted for 80% isopropanol for washes (*see* **Note 7**).
2. Elute the viral DNA in 50 µL of prewarmed (80°C) water.
3. Store the purified viral DNA samples at –20°C if necessary.

3.4. Real-Time PCR

1. Prepare a master mix for the real-time PCR reaction as listed below. The volumes shown below are for a single reaction. Prepare enough master mix to run each reaction in duplicate (*see* **Note 8**):
   a. 25 µL 2X SYBR® Green PCR Master Mix.
   b. 2 µL Forward primer (1.25 µ*M* working stock).
   c. 2 µL Reverse primer (1.25 µ*M* working stock).
   d. 14 µL RNase- and DNase-free water.
2. Dispense 43 µL of the real-time PCR master mix into each reaction vessel (*see* **Note 9**).
3. Prepare a 1:10 and 1:100 dilution of each unknown to be analyzed (*see* **Note 10**).
4. Add 7 µL of water (No Template Control, NTC), standard DNA (pJW228), or diluted unknown to each reaction vessel. Run each sample in duplicate.
5. Seal the vessels with the appropriate cover (*see* **Note 11**). Use the cap installing tool to tightly secure the caps.
6. Gently shake the samples.
7. Centrifuge the samples (1 min at approx 250*g*) to bring each sample to the bottom of the well.
8. Place the reaction vessels in the thermal cycler and then program the ABI Prism 7700 Sequence Detection System software (SDS software) as described below:
   a. Denature the DNA for 10 min at 95°C for one cycle.
   b. Incubate for 15 s at 95°C.
   c. Incubate for 1 min at 60°C.
9. Repeat **steps 8b** and **8c** for 40 total cycles.
10. At the end of the run, adjust the baseline start and stop values on the amplification plot if necessary (*see* **Note 12**). An amplification plot of the plasmid standard samples is shown (**Fig. 1A**).

Fig. 1. (*opposite page*) Typical amplification plot and standard curve for Ad5 real-time polymerase chain reaction (PCR) assay standards. (**A**) An amplification plot for replicates of the plasmid standard samples is shown. The change in fluorescence (Δ*Rn*)

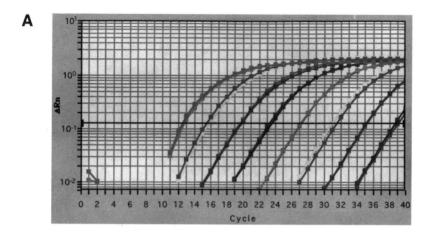

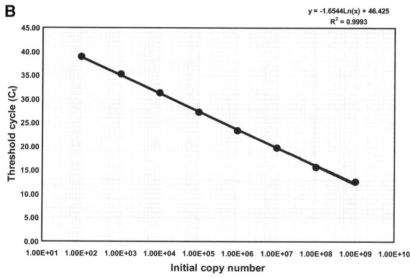

Fig. 1. *(continued from opposite page)* is plotted against the PCR cycle number. Rn, the normalized reporter signal, is defined as the SYBR Green intensity divided by the intensity of ROX, an internal fluorescent dye that remains unchanged during PCR. The thick black bar represents the threshold and is used to determine the C_t values. Samples with higher input copy number cross the threshold at lower cycle number and thus are situated toward the left of the graph. In this case the leftmost curves represent amplification of the samples that contained 1×10^9 copies of the plasmid pJW228. **(B)** A standard curve was generated by plotting the threshold cycle (C_t) against the initial plasmid copy number for each of the standards. The quantitative range of the standard curve was found to be between 1×10^2 and 1×10^9 copies. The resultant line equation can be used to determine the initial copy numbers for unknown samples, based on their C_t values.

11. Adjust the threshold so that it falls within the linear portion of the amplification curve when the y axis is in log scale. The point at which the amplification plot of each sample crosses the threshold bar is defined as the threshold cycle (C_t) value.
12. The SDS software will automatically generate a standard curve by plotting the C_t of each standard versus the copy number. A standard curve generated for the plasmid standards is shown (**Fig. 1B**). The SDS software will also calculate the initial copy number of each unknown based on its C_t value and the standard curve (*see* **Note 13**).

3.5. Titer Determination

1. From the initial copy numbers generated above, the titer (real-time PCR units per mL) of each unknown sample can now be calculated as shown below (*see* **Note 14**):

$$\text{Titer} = \frac{\text{Copy Number}}{7\mu\text{L}} \times 10 \times \text{Dilution Factor} \times \frac{1 \times 10^3\,\mu\text{L}}{\text{mL}}$$

2. This titer can be correlated with a standard titration technique used in your laboratory, such as the plaque assay or the endpoint dilution assay. Plot the titers generated by real-time PCR against the titers generated by a standard assay.
3. Using the correlation between the two titration methods, the real-time PCR method can be used to quickly estimate the titer of a stock that has not yet been titered by the standard assay. For example, we have found that titers determined by this real-time PCR assay correlate with infectious titers determined by plaque assay *(6)* (**Fig. 2**; *see* **Note 15**) and endpoint dilution assay *(7)*. Particle titers determined by absorbance at 260 nm were found to correlate with neither real-time PCR titers nor plaque assay titers. These observations emphasize the unique ability of this real-time PCR assay to correlate with infectious titers.

4. Notes

1. This assay has not been used with other real-time PCR systems. The ABI 7500 has an emulation mode that will adapt the protocol for this machine. For use on other machines, the assay may need to be optimized.
2. Standard precautions for limiting contamination should be taken such as working in a UV hood, using dedicated PCR pipettors and filter-plugged pipet tips, and working with positive control samples last. Consider setting up the PCR in a different laboratory to limit possible exposure to contaminating DNA. Pipettors should be cleaned after each use with a decontaminating solution such as ELIMINase and/or briefly placed under a UV light source.
3. Prior to use, dilute an appropriate amount of enzyme to 0.5 U/µL by mixing one part RQ1 DNase with one part water.
4. The use of a heat block for all incubations reduces the possibility of cross-contamination.
5. After the DNase digestion has begun, prepare a 75°C water bath or heat block.

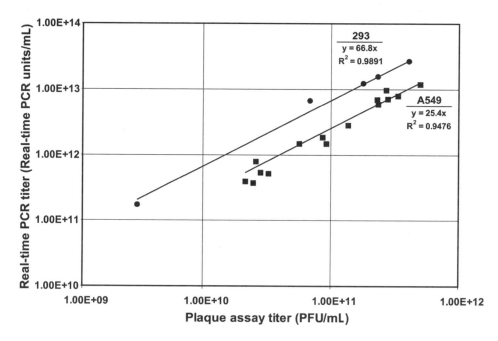

Fig. 2. Correlation of real-time polymerase chain reaction (PCR) titers and plaque assay titers. The titer of each stock as determined by real-time PCR was plotted against the respective plaque assay titer. Real-time PCR titers correlated with plaque assay titers determined on A549 cells with a correlation coefficient of 0.9476 and with titers determined on 293 cells with a correlation coefficient of 0.9891. The ratio of real-time PCR titer to plaque assay titer was approx 25 for stocks plaque assayed on A549 cells and 67 for stocks plaque assayed on 293 cells.

6. After the Proteinase K digestion has begun, prepare prewarmed (80°C) water (50 μL per sample, plus extra) for elution after DNA purification.
7. We used the "Protocol for DNA Purification Without a Vacuum Manifold," in which each sample is processed in a disposable 3-mL Luer-Lok syringe.
8. In addition to the unknown samples and the range of plasmid standards, run several replicates of water (no template control, NTC). With each run, you may also include a few viral DNA samples from stocks that have been titered previously by plaque assay and real-time PCR.
9. Use either a MicroAmp optical 96-well reaction plate (if close to 96 samples will be assayed) or MicroAmp optical tubes.
10. Typically, the 1:100 dilution of a virus stock will fall within the range of the standard curve. If the titer is expected to be low, the 1:10 dilution may be used.
11. The optical caps may be used with either the optical 96-well reaction plate or the optical tubes. The optical adhesive cover may only be used with the 96-well reaction plate.

12. The baseline start and stop values are set by default at cycles 3 and 15, respectively. To determine whether these values should be adjusted, double-click on the *y*-axis and change the plot to linear view. If the first sample to be amplified emerges after cycle 15, then no changes are needed. If, however, the first sample emerges prior to cycle 15, set the baseline stop value to fall just prior to the cycle at which the first sample is amplified. Click OK and return the plot to log view.

13. Alternatively, the standard curve and resultant C_t values for unknowns can be generated in Microsoft Excel. This method gives the researcher more control over generation of the standard curve.

14. The 7-µL portion of the formula is from the volume of diluted viral DNA that was added to each well for real-time PCR. The multiplication by 10 reflects the fact that the viral DNA was eluted in 50-µL volume and the virus was originally obtained in a volume of 5 µL. The dilution factor will typically be 10 or 100, depending on which dilution was assayed.

15. Interestingly, we found that the correlation between titers generated by real-time PCR and titers determined by plaque assay depended on the cell line used for plaque assay. Titers from plaque assays performed on A549 and 293 both correlated with real-time PCR titers, but the ratio of real-time PCR units to plaque-forming units was different on the two different cell lines.

References

1. Tollefson, A. E., Scaria, A., Hermiston, T. W., Ryerse, J. S., Wold, L. J., and Wold, W. S. M. (1996) The adenovirus death protein (E3-11.6K) is required at very late stages of infection for efficient cell lysis and release of adenovirus from infected cells. *J. Virol.* **70,** 2296–2306.

2. Mittereder, N., March, K. L., and Trapnell, B. C. (1996) Evaluation of the concentration and bioactivity of adenovirus vectors for gene therapy. *J. Virol.* **70,** 7498–7509.

3. Heim, A., Ebnet, C., Harste, G., and Pring-Akerblom, P. (2003) Rapid and quantitative detection of human adenovirus DNA by real-time PCR. *J. Med. Virol.* **70,** 228–239.

4. Ma, L., Bluyssen, H. A. R., De Raeymaeker, M. D., et al. (2001) Rapid determination of adenoviral vector titers by quantitative real-time PCR. *J. Virol. Methods* **93,** 181–188.

5. Watanabe, M., Kohdera, U., Kino, M., Haruta, T., Nukuzuma, S., Suga, T., Akiyoshi, K., Ito, M., Suga, S., and Komada, Y. (2005) Detection of adenovirus DNA in clinical samples by SYBR Green real-time polymerase chain reaction assay. *Pediatr. Int.* **47,** 286–291.

6. Tollefson, A. E., Hermiston, T. W., and Wold, W. S. M. (1998) Preparation and titration of CsCl-banded adenovirus stocks, in *Adenovirus Methods and Protocols* (Wold, W. S. M., ed.), The Humana Press, Totowa, NJ, pp. 1–9.

7. Condit, R. C. (2001) Principles of virology, in *Fields Virology* (Knipe, D. M., Howley, P. M., Griffin, D. E., Martin, M. A., Lamb, R. A., Roizman, B., and Straus, S. E., eds.), Lippincott Williams and Wilkins, Philadelphia, pp. 19–51.

15

Detection and Quantitation of Subgroup C Adenovirus DNA in Human Tissue Samples by Real-Time PCR

C. T. Garnett, Ching-I Pao, and Linda R. Gooding

Summary

Advances in amplification techniques have revolutionized the ability to detect viruses both quantitatively and qualitatively and to study viral load. Real-time polymerase chain reaction (PCR) amplification depends on the ability to detect and quantify a fluorescent reporter molecule whose signal increases in proportion to the amount of amplification product generated. Recent advances have been made by using probes, such as TaqMan probes, to detect amplified products. Use of these probes offers confirmation of specificity of the PCR product. Here we describe a sensitive real-time PCR assay to quantify subgroup C adenoviral DNA in human lymphocytes derived from mucosal tissues removed in routine tonsillectomy or adenoidectomy. This chapter will describe in detail the methods used for these analyses.

Key Words: Subgroup C human adenoviral DNA; quantitative PCR; real-time PCR; TaqMan probes; absolute DNA copy number; normalization; adenovirus serotype determination.

1. Introduction

Adenovirus persists at low levels that are difficult to quantify using standard PCR techniques *(1–5)*. Although the technology to detect PCR products in real time has been available for more than 10 yr *(6)*, the use of this technology has increased dramatically in the past 5 yr. Real-time PCR has been successfully used for the absolute quantification of subgroup C adenoviral DNA from human tissues *(7,8)*. This assay amplifies a 285-base-pair fragment from a region of the hexon gene, which is relatively conserved among subgroup C serotypes but varies considerably in serotypes from other adenovirus subgroups. Using this assay, subgroup C DNA has been detected in Ficoll-purified lymphocytes from 107 of 180 tissue specimens tested (59%) *(7)*. The levels of

From: *Methods in Molecular Medicine, Vol. 130:*
Adenovirus Methods and Protocols, Second Edition, vol. 1:
Adenoviruses Ad Vectors, Quantitation, and Animal Models
Edited by: W. S. M. Wold and A. E. Tollefson © Humana Press Inc., Totowa, NJ

DNA varied from the lower limit of detection to 2×10^6 copies of the adenoviral genome/10^7 lymphocytes, depending on the donor. DNA from adenovirus serotypes 1, 2, and 5 were detected in tonsil and adenoid tissues, whereas serotype 6 was not. Serotype 6 was readily detected in tissues from intussusception patients *(8)*, indicating that this serotype can be detected by this method. With this method one can determine relative viral DNA copy numbers in a sample by relating the PCR signal to an external standard curve. Other advantages of real-time PCR include enhanced specificity of product, reduced variation, and the lack of a requirement for post-PCR manipulations. The high specificity of this method is a result of complementarity between the set of primers, the internal probe and the target sequence. With the use of the TaqMan probe, a fluorescent signal will be generated only if the probe is annealed to the target sequence during amplification. The detection of subgroup C adenoviral DNA in human lymphocytes will be used to illustrate the method. The steps described below outline the following:

1. The preparation of single-cell suspensions from human tissues.
2. The isolation of lymphocytes.
3. The preparation of DNA from cell lysates.
4. The real-time PCR assay setup.
5. Subgroup C serotype determination.

The real-time PCR method described here has been undergoing minor modifications since its original publication *(7)*. Hence, the method described here is the current (2005) version.

2. Materials

1. HH5 medium: *N*-2-hydroxyethylpiperazine *N*'-2-ethanesulfonic acid (HEPES)-buffered Hank's balanced salt solution (HH), 5% fetal calf serum (FCS), 0.05 mg/mL gentamicin, 1% antibiotic–antimycotic mixture (AA, Gibco).
2. Microsurgical scissors and forceps soaked first in 10% Clorox for 30 min, then soaked in 70% ethanol for 30 min to sterilize and to destroy DNA.
3. Sterile 100-mesh stainless steel wire screen cut to fit tightly into 60-mm tissue culture dishes.
4. Sterile 10-cc syringe plungers.
5. Cell-freezing medium: 90% fetal calf serum (FCS), 10% dimethyl sulfoxide (DMSO).
6. Ficoll Type 400 (Sigma).
7. Phosphate-buffered saline (PBS).
8. Isotonic lysis buffer: 10 mM Tris-HCl, pH 8.3, 1.5 mM MgCl$_2$, 75 mM KCl, 0.5% NP-40, 0.5% Tween-20, 0.5 mg/mL Proteinase K (freshly added).
9. PCR-grade water.
10. Commercial adenovirus type 2 DNA: Ad2 DNA (Gibco).
11. UV lamp.

12. Qiagen 10X PCR buffer.
13. Qiagen Hotstart Taq polymerase.
14. 50 µ*M* of forward adenovirus hexon primer: (forward-5' GCCATTACCTTTGA CTCTTCTGT 3').
15. 50 µ*M* of reverse adenovirus hexon primer: (reverse-5' GCTGTTGGTAGTCC TTGTATTTAGTATC 3').
16. TaqMan adenovirus probe (light sensitive): ([FAM] 5' AGAAACTTCCAGCCC ATGAGCCG 3' [BHQ1]).
17. 96-Well iCycler optical PCR plate and plate optical sealer tape.
18. Real-time thermocycler (*see* **Note 1**).
19. Agarose gels.
20. Qiagen PCR purification kit.

3. Methods

Before beginning the procedure it is strongly recommended that different rooms and/or dedicated equipment be used for: (1) DNA purification and processing, (2) PCR setup, and (3) post-PCR gel analysis in order to avoid sample contamination.

3.1. Human Tissue Sample Processing

1. Collect surgically removed human tonsil and adenoid tissue in sterile HH5 (*see* **Note 2**).
2. Place the tissue in a 60-mm dish containing sterile wire mesh and 10 mL of HH5 medium (*see* **Note 3**) and mince tissue into smaller manageable pieces using sterile scissors.
3. Gently push the tissue pieces through the stainless steel wire screen using the rubber plunger from a 10-cc sterile syringe. This will produce a single-cell suspension and remove any connective tissue.
4. Collect the cell-containing medium that has passed through the wire screen, and transfer it to a sterile 50-mL centrifuge tube.
5. Rinse the screen with an additional 10 mL of HH5 medium and transfer the wash to the 50-mL tube.
6. Centrifuge and wash the cells once in medium; then count and store in aliquots of $1-2 \times 10^8$ cells per mL of cell-freezing medium in the vapor phase of the liquid nitrogen freezer (*see* **Note 4**).

3.2. Lymphocyte Purification and DNA Isolation

DNA preparations are made from single-cell suspensions of Ficoll-purified lymphocytes. This method of DNA isolation is designed to minimize DNA loss that can result from additional manipulations of the sample such as extraction and multiple wash steps. Accordingly, it is not intended to produce the most pure DNA preparation. Lymphocytes were isolated for our studies, but other cell types can be used as long as single-cell suspensions are prepared and accurate cell counts are achieved.

1. Prepare lymphocytes for DNA isolation by thawing an aliquot of cells (**Subheading 3.1., step 6**), diluting to 10 mL with HH5, pelleting the cells at 180g for 5 min, and resuspending cells in 3 mL HH5 in a 15-mL conical centrifuge tube.
2. Underlay the cell suspension with 3 mL of Ficoll (*see* **Note 5**).
3. Centrifuge the cell suspension/Ficoll gradient for 25 min at 700g in a 20–25°C centrifuge, with no brake (*see* **Note 6**).
4. Following centrifugation, a cloudy layer of cells should be visible suspended between the medium and the Ficoll. Collect these cells and wash them in 25 mL of PBS containing 2% FCS (*see* **Note 7**).
5. Count the cells. DNA is extracted from aliquots of 5×10^6 purified lymphocytes. Pellet cells in a 1.5-mL microcentrifuge tube by centrifugation at 2500g in an Eppendorf 5415C microfuge for 5 min. Aspirate the supernatant and retain the cell pellet.
6. Add 200 µL of isotonic lysis buffer (*see* **Note 8**) directly to the cell pellet. Mix samples and incubate at 55°C for 12 h essentially as described by Babcock et al *(9)*.
7. Following incubation at 55°C, inactivate Proteinase K by incubation at 95°C for 15 min in a water bath (*see* **Note 9**).
8. Vortex samples vigorously for 1–2 min, spin in a microfuge for 5 min, and add 5 µL directly into the PCR reaction described below.
9. Use DNA preparations immediately or stored at –20°C until use.

3.3. Real-Time Quantitative PCR for Adenovirus Hexon DNA

Quantitative analysis of subgroup C adenovirus DNA from these cellular DNA preparations can now be performed using real-time PCR *(8)*. To assess the presence of adenovirus subgroup C DNA, a sensitive real-time PCR assay was developed using adenovirus hexon-specific primers and a TaqMan probe. Primers were selected from a region that is highly conserved among subgroup C viruses, but is significantly divergent among other subgroups. Primers were modified from those originally described by Pring-Akerblom et al. *(10)* to facilitate amplification of a conserved region of the subgroup C adenovirus hexon gene (nucleotides 21005–21290 of adenovirus type 5 (Ad5), GeneBank accession number AY339865). Primer and TaqMan probe sequences are presented in **Fig. 1A**. This reaction produces a 285-bp PCR product, and, as expected, the primers are able to amplify all four of the subgroup C viruses (Ad1, Ad2, Ad5, and Ad6) but do not amplify representatives from the other subgroups (**Fig. 1B**) (*see* **Note 10**). Serial dilutions (from 5×10^7 to 1 copy) of Ad2 DNA are included in each run to generate a standard curve for quantitative assessment of donor adenovirus DNA. This assay is able to detect five copies of the adenovirus genome (*see* **Note 11**). The range of the assay allows quantitation over at least 7 orders of magnitude (**Fig. 1C**).

A

Group	Primer (F)	5' > 3'																						
		G	C	C	A	T	T	A	C	C	T	T	T	G	A	C	T	C	T	T	C	T	G	T
C	Ad1	-	-	-	-	-	-	C	-	-	-	-	-	-	-	-	-	-	-	-	-	-	-	-
C	Ad2	-	-	-	-	-	-	-	-	-	-	-	-	-	-	-	-	-	-	-	-	-	-	-
C	Ad5	-	-	-	-	-	-	-	-	-	-	-	-	-	-	-	-	-	-	-	-	-	-	-
C	Ad6	-	-	-	-	-	-	-	-	-	T	-	-	-	-	-	-	-	-	-	-	-	-	-
A	Ad12	T	-	A	-	-	-	C	-	T	G	-	-	-	-	-	-	-	-	C	-	-	C	-
B	Ad7	T	-	-	-	-	-	C	-	T	G	-	-	-	-	-	-	-	-	C	-	-	A	-
B	Ad16	T	-	-	-	-	-	C	-	T	G	-	-	-	-	-	-	-	-	A	-	-	-	-
D	Ad48	T	-	-	-	-	-	C	-	T	G	-	-	C	-	-	-	-	-	C	-	-	G	-
E	Ad4	T	-	-	-	-	-	C	-	-	-	-	-	C	-	-	-	-	-	C	-	-	-	-
F	Ad41	T	-	-	-	-	-	C	-	T	G	-	-	-	-	-	-	-	-	C	-	-	-	-

Group	Primer (R)		C	C	T	G	T	T	G	G	T	A	G	T	C	C	T	T	G	T	A	T	T	T	A	G	T	A	T	C
C	Ad1		-	-	-	-	-	-	-	-	-	-	-	-	-	-	-	-	-	-	-	-	-	-	-	G	-	-	-	-	-
C	Ad2		-	-	-	-	C	-	-	A	-	-	C	-	-	-	-	-	-	-	-	-	-	-	-	-	-	-	-	-	-
C	Ad5		-	-	-	-	-	-	-	-	-	-	-	-	-	-	-	-	-	-	-	-	-	-	-	-	-	-	-	-	-
C	Ad6		-	-	-	-	C	-	-	A	-	-	A	-	-	T	-	-	-	-	-	-	-	-	-	-	-	-	-	-	-
A	Ad12		-	T	-	T	-	-	T	-	-	-	-	-	T	-	-	-	A	-	-	-	-	C	T	-	-	-	G	G	T
B	Ad7		-	G	G	C	-	-	T	-	-	-	-	-	A	G	-	-	-	A	-	-	-	A	C	C	-	-			
B	Ad16		-	G	G	C	-	-	T	-	-	-	-	-	A	G	-	-	-	A	-	-	-	A	C	C	-	-			
D	Ad48		-	G	G	C	C	-	T	-	-	-	-	-	-	-	-	-	-	G	-	-	G	A	-	C	-	-			
E	Ad4		-	G	G	C	C	-	-	-	-	-	-	-	-	-	-	-	-	G	-	-	-	A	C	C	-	-			
F	Ad41		-	A	-	T	C	-	-	A	-	-	T	-	-	-	-	-	-	-	G	G	-	G	-	-	-	-	G	T	

Group	Probe			A	G	A	A	A	C	T	T	C	C	A	G	C	C	C	A	T	G	A	G	C	C	G
C	Ad1			-	-	-	-	-	-	-	-	-	-	-	-	-	-	-	-	-	-	-	-	-	-	-
C	Ad2			-	-	-	-	-	-	-	-	-	-	-	-	-	-	-	-	-	-	-	-	-	-	-
C	Ad5			-	-	-	-	-	-	-	-	-	-	-	-	-	-	-	-	-	-	-	-	-	-	-
C	Ad6			-	-	-	-	-	-	-	-	-	-	-	-	-	-	-	-	-	-	-	-	-	-	-
A	Ad12			-	-	-	-	-	-	-	T	-	-	-	-	-	-	-	-	-	-	-	-	T	A	-
B	Ad7			-	-	-	-	-	-	-	-	-	-	-	-	T	-	-	-	-	-	-	-	A	-	
B	Ad16			-	-	-	-	-	-	-	-	-	-	-	-	T	-	-	-	-	-	-	-	A	-	
D	Ad48			C	-	C	-	-	-	-	-	-	-	-	-	-	-	-	-	-	-	-	-	A	-	
E	Ad4			C	-	C	-	-	-	-	-	-	-	-	-	-	-	-	-	-	-	-	-	-	-	
F	Ad41			C	-	-	-	-	-	-	-	-	-	-	A	-	-	-	-	-	T	-	-		

B

		Group C						Group					
								B	E	B	D	A	F
M	1	2	5	6	-	+	3	4	7	9	31	41	M

Fig. 1. (**A**) Nucleotide sequences of primers and TaqMan probe are to the hexon region of adenovirus. (**B**) The subgroup C viruses as well as representatives from each of the other human adenovirus subgroups were amplified by real-time PCR using the hexon primers, and the products were run on an ethidium bromide-stained 1.8% agarose gel. Numbers indicate the adenovirus serotype tested. M, marker; −, negative water control; +, Ad2-positive control DNA.

3.3.1. Preparation of Viral DNA Standards

In addition to your samples of interest, each PCR run will need to include a series of wells containing viral DNA standard dilutions (*see* **Note 12**).

1. Determine the concentration of standard Ad genomic DNA (Gibco) and convert to copy numbers (molecules) of viral DNA (*see* **Note 13**).
2. Prepare 10-fold serial dilutions of Ad2 genomic DNA, ranging from 5×10^7 copies/5 µL to 5 copies/5 µL in PCR-grade water containing 50 ng/mL yeast tRNA (Ambion).
3. Aliquot diluted standards and store at –20°C until PCR setup (*see* **Note 14**).

3.3.2. Preparation of Primers

Prepare 50 µ*M* (100X) concentrations of each primer in PCR-grade water, divide into 50 µL aliquots, and store at –20°C (*see* **Note 15**).

3.3.3. PCR Setup

1. Decontaminate the PCR work station by turning on a UV light for 15 min prior to setting up any PCR reaction (*see* **Note 16**).
2. Thaw lymphocyte lysate and adenovirus DNA standards on ice, mix thoroughly by vortexing, and distribute each sample (5 µL/well) into a 96-well iCycler PCR plate. Keep the plate on ice during the PCR setup. Samples are generally run in triplicate for both unknowns and standards. Run the five copies per well standard in replicates of five.
3. Prepare the PCR master mix just before use with the following final concentrations: Qiagen 1X PCR buffer, 2.35 m*M* MgCl$_2$, 0.2 m*M* dNTPs, 0.5 µ*M* of each primer, 2.5 U Qiagen Hotstart Taq polymerase, and 0.1 µ*M* of probe. Use PCR-grade water to balance the total volume to 45 µL per reaction. Mix thoroughly and place 45-µL aliquots into 96-well PCR plate (*see* **Note 17**).
4. Cover the PCR plate with iCycler optical tape and centrifuge briefly at 300*g* to ensure that none of the PCR reaction solution is stuck to the tape or the well walls.
5. The PCR plate is run in a Bio-Rad iCycler thermocycler at: 1 cycle of 95°C × 15 min, followed by 50 cycles of 95°C, 15 s, 55°C, 1 min.

3.3.4. Data Analysis

With known amounts of input copy number, the target gene can be quantified in the unknown samples. To quantify the results we included serial dilutions of purified Ad2 DNA in each PCR run. To make a correlation between the initial template concentration and the real-time detection curve, a point has to be determined where the fluorescence signal exceeds the average background signal. This point is referred to as the cycle threshold (C_T) value (*see* **Note 18**). The C_T is reported as a cycle number at this point and decreases linearly with

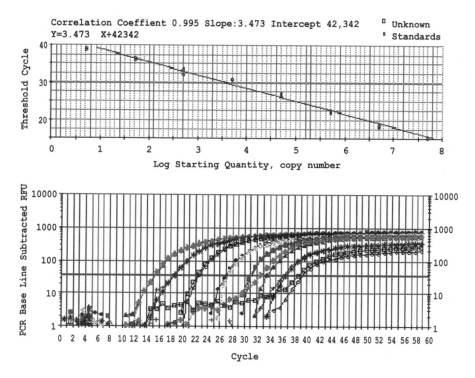

Fig. 2. (**A**) Purified Ad2 DNA was serially 10-fold diluted, and two replicates of each dilution (from 5×10^7 to 5 copies) were tested. The threshold cycle values of the standard dilution are plotted against the $\log_{10}$ of the starting copy number. The equation of the line gives a correlation coefficient higher than 0.990, and the slope of the line is greater than –3.8. (**B**) The fluorescence intensity collected in real time for each sample was plotted against the number of PCR cycles. The horizontal orange line indicates the fluorescence cycle threshold (C_T) setting, which is set at 10 standard deviations above the baseline emission. RFU, relative fluorescent unit.

increasing input target quantity. By plotting C_T values against the known input copy number, a standard curve is generated with linear range covering 7–8 log units (*see* **Note 19**). The C_T value (or cycle number where the fluorescence of the unknown crosses threshold) of the unknown sample can then be correlated to the copy number of the standard with a corresponding C_T value.

The final quantity of viral DNA must be reported relative to some biological standard such as cell number (*see* **Fig. 2**) or weight of tissue or volume of serum. For this reason, accurate input information, such as cell counts, is essential. Alternatively, one can normalize samples to an external reference gene (*see* **Note 20**).

3.3.5. Subgroup C Serotype Determination

The real-time PCR assay described above efficiently detects all subgroup C adenoviruses. Nonetheless, there are small differences in sequence of the PCR product among the four subgroup C serotypes (Ad1, 2, 5, and 6). Thus the dominant serotype present in a clinical sample can be determined by sequencing the PCR product (*see* **Note 21**).

1. Run the real-time PCR hexon products on a 1.8% agarose gel.
2. Cut out the PCR bands of interest using a fresh razor blade for each sample.
3. Purify the gel slice using a Qiagen gel extraction kit according to manufacturer's instructions.
4. Elute the purified DNA in sterile water and sequence using the forward real-time PCR hexon primer (*see* **Note 22**).

Virus serotype is determined by sequence comparison between the PCR product sequence and known nucleotide sequences for subgroup C adenoviruses published in GenBank by the National Center for Biotechnology Information (NCBI). Serotypes are identified by conserved nucleotide base changes (**Fig. 3**).

4. Notes

1. We used an iCycler iQ from Bio-Rad Instruments (Hercules, CA). It has an optical module that can be connected in the thermal cycler, measuring fluorescence emission during the course of PCR amplification. Ninety-six wells are tracked simultaneously, thereby providing a very fast assay.
2. Human tonsil and adenoid tissues were used in our studies, but other human tissue types could be similarly analyzed. In addition, alternate media may be substituted for the HH5 medium used here as long as the antibiotics are maintained. This is particularly important if tonsil and adenoid tissues are used because they come from the nonsterile environment of the throat.
3. If using human tonsil specimens, it is best to prepare each tonsil in a separate dish with wire mesh, as they are rather large pieces of tissue.
4. Samples can be used immediately at this point, without freezing, if desired. Use of cells prior to freezing will result in higher cell yields and viability.
5. Care must be taken to avoid mixing the cell and Ficoll layers prior to centrifugation. Ficoll gradients, which enrich lymphocytes and remove dead cells, may not be necessary for other cell types examined.
6. Having the brake on for the centrifuge can cause the gradient between the Ficoll and the medium to be disrupted and to mix as the centrifuge slows down.
7. To collect lymphocytes from the gradient, place the tip of a transfer pipet near the cloudy cell layer and collect the cells. It is acceptable to collect some of the medium as well as the Ficoll in addition to the cells because this will be removed in the subsequent wash step.

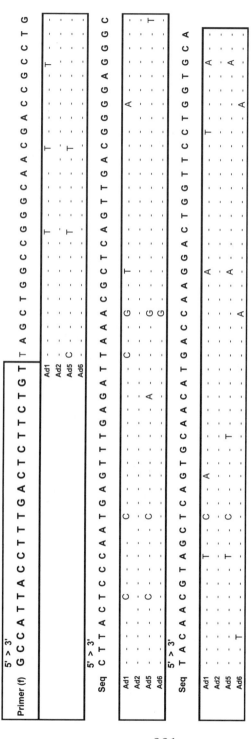

Fig. 3. Distribution and quantitation of adenovirus DNA isolated per 10^7 lymphocytes in 42 tonsil and adenoid samples from 35 donors. Real-time polymerase chain reaction (PCR) was performed on DNA purified from Ficoll-purified lymphocytes from tonsils (T) and adenoids (A). Cellular input DNA amounts were normalized to quantities of GAPDH between samples being compared. $N = 2$–5 individual donors.

8. Lysis buffer can be stored at room temperature or at 4°C with all components except Proteinase K. Store the Proteinase K at –20°C and add it to lysis buffer just before use. We have determined that volumes of lysis buffer less than 200 µL per 5×10^6 lymphocytes are insufficient to ensure that all cells are completely lysed and that the DNA has been released.

9. It is best to use 1.5-mL screw-cap tubes for this step because snap caps may pop open when heated to 95°C at this step.

10. The faint band seen in the Ad4 lane is nonspecific because the hexon TaqMan probe does not hybridize with this product and no fluorescence signal is detected (data not shown).

11. The precision for high copy numbers is better that for low copy numbers. At low copy numbers, the statistical variation resulting from sampling becomes an important factor *(11)*.

12. Preparing standards in an area separate from the area where tissue samples and DNA preparations are done will minimize the risk of sample contamination.

13. The accuracy of quantitation depends on the accuracy of the standards. For this reason, a commercial preparation of adenoviral DNA is preferred. Ad2 DNA is 35,937 bp, and the concentration of commercial DNA can be converted to copy number using the formula: 1 µg of 1000 bp DNA = 9.1×10^{11} molecules. Alternatively, the concentration of the DNA sample can be measured spectrophotometrically at 260 nm and converted to molecules in the same manner.

14. It is helpful to make a large volume (1 mL) of each standard dilution, then divide into small aliquots and store at –20°C *(12)*. This will help ensure that experiments done at different times have reproducible standards to optimize comparison of experiments run on different days. To minimize interassay variability, all samples analyzed from a single experiment should be performed with the same dilution series of standard.

15. Aliquoting of primers and probes is suggested to avoid multiple freeze–thaw cycles, which can result in reduced sensitivity of the reagents. This is especially true of the probe. Care should be taken to aliquot the probe immediately upon receipt. Store the probe at –20°C, protected from light.

16. For our studies, a plastic hood was set up in an area designated for PCR setup only. Positive displacement pipets were stored in this hood and used only for PCR setup. The use of positive displacement pipets for plating positive control standards as well as test samples can help avoid sample cross-contamination.

17. The results from different runs may have to be compared when performing experiments with large numbers of samples, and precautions should be taken to minimize interassay variability. This comparison can be achieved by preparation of one universal PCR mixture with subsequent running of samples from a single experimental setup using this same master mix *(12)*. Experimental samples should be interspersed with blank or negative controls on the plate. Of course, a positive signal detected in a negative control sample indicates contamination, in which case the entire assay should be discarded.

18. This C_T value is arbitrarily chosen; for this method we chose 10 standard deviations above the no-template background fluorescence. This can be different for each individual experiment if necessary.

19. When the threshold cycle values of the standard dilutions are plotted against the $\log_{10}$ of the starting copy number, correlation coefficient values should be higher than 0.98 and the slope of the line greater than –3.8, indicating high amplification efficiency according to Bio-Rad, the manufacturer of the iCycler. If the PCR reaction is 100% efficient, and there is a doubling of product with each cycle; it takes approx 3.3 cycles for a 10-fold (dilution of standards) increase.

20. Normalization to a reference gene is currently one of the most acceptable methods to correct for minor variations as a result of differences in input DNA amount or in reaction efficiencies. If this is desired, a second set of duplicate wells containing your unknowns can also be included for amplification of a gene used for normalization of samples. All samples can be tested for the gene glyceraldehyde-3-phosphate dehydrogenase (GAPDH) to normalize variation between samples caused by differences in cell counts or pipetting. Amplify GAPDH using real-time PCR and using the PCR master mix recipe as previously described with the following primers and probe:
    a. Sense primer (5'AAATGAATGGGCAGCCGTTA 3').
    b. Antisense primer (5'TAGCCTCGCTCCACCTGACT 3').
    c. TaqMan probe (5'-FAM CCTGCCGGTGACTAACCCTGCGCTCCT BHQ-3'). This reaction produces a 105-bp PCR product. Samples with more DNA loaded will have a lower C_T value for GAPDH than those with less DNA. From this information, a ratio can be calculated and all samples normalized to the same GAPDH amounts.

21. If this step is performed, extreme caution should be taken to avoid contamination of tissue preparation and PCR setup areas with the PCR product. This can result in sample contamination and false-positive results.

22. Alternately, the reverse primer can be used for sequencing, but the resulting 5' to 3' sequence will be the reverse complement. Sequencing of the PCR product was routinely performed for us by a core facility that requests that DNA preparations be supplied in water. DNA samples should be prepared in the buffer requested by those who will perform the sequencing.

References

1. Lavery, D., Fu, S. M., Lufkin, T., and Chen-Kiang, S. (1987) Productive infection of cultured human lymphoid cells by adenovirus. *J. Virol.* **61,** 1466–1472.
2. Macek, V., Sorli, J., Kopriva, S., and Marin, J. (1994) Persistent adenoviral infection and chronic airway obstruction in children. *Am. J. Respir. Crit. Care Med.* **150,** 7–10.
3. Marin, J., Jeler-Kacar, D., Levstek, V., and Macek, V. (2000) Persistence of viruses in upper respiratory tract of children with asthma. *J. Infect.* **41,** 69–72.
4. Neumann, R., Genersch, E., and Eggers, H. J. (1987) Detection of adenovirus nucleic acid sequences in human tonsils in the absence of infectious virus. *Virus Res.* **7,** 93–97.

5. Silver, L., and Anderson, C. W. (1988) Interaction of human adenovirus serotype 2 with human lymphoid cells. *Virology* **165,** 377–387.
6. Heid, C. A., Stevens, J., Livak, K. J., and Williams, P. M. (1996) Real time quantitative PCR. *Genome Res.* **6,** 986–994.
7. Garnett, C. T., Erdman, D., Xu, W., and Gooding, L. R. (2002) Prevalence and quantitation of species C adenovirus DNA in human mucosal lymphocytes. *J. Virol.* **76,** 10,608–10,616.
8. Guarner, J., de Leon-Bojorge, B., Lopez-Corella, E., et al. (2003) Intestinal intussusception associated with adenovirus infection in Mexican children. *Am. J. Clin. Pathol.* **120,** 845–850.
9. Babcock, G. J., Decker, L. L., Volk, M., and Thorley-Lawson, D. A. (1998) EBV persistence in memory B cells in vivo. *Immunity* **9,** 395–404.
10. Pring-Akerblom, P., Trijssenaar, F. E., Adrian, T., and Hoyer, H. (1999) Multiplex polymerase chain reaction for subgenus-specific detection of human adenoviruses in clinical samples. *J. Med. Virol.* **58,** 87–92.
11. Niesters, H. G. (2001) Quantitation of viral load using real-time amplification techniques. *Methods* **25,** 419–429.
12. Giulietti, A., Overbergh, L., Valckx, D., Decallonne, B., Bouillon, R., and Mathieu, C. (2001) An overview of real-time quantitative PCR: applications to quantify cytokine gene expression. *Methods* **25,** 386–401.

16

Flow Cytometric Detection of Adenoviruses and Intracellular Adenovirus Proteins

Graham Bottley, John R. Holt, Nicola J. James, and G. Eric Blair

Summary

Precise and simple assay of purified and crude preparations of human adenoviruses is essential for basic and gene therapy research. Previous bioassays used to quantitate adenoviruses (such as the plaque assay or fluorescent focus assay) are time-consuming and subjective in their interpretation. Here we describe a flow cytometric method that eliminates these disadvantages and provides a quantitative and reliable method of focus-forming unit (FFU) assay.

Key Words: Adenovirus; FFU; flow cytometry; Hexon protein; immunofluorescence; adenovirus bioassay.

1. Introduction

1.1. Assay of Human Adenoviruses

Assay of the infectivity of human adenoviruses is an essential procedure in most experimental work on virus–host cell systems as well as for the increasing use of recombinant adenovirus vectors in clinical gene therapy protocols. Methods used for quantitative assay of adenovirus infectivity have relied largely on the plaque assay *(1,2)* which provides a reliable bioassay for most, but not all, of the adenovirus serotypes. The requirement for the adenovirus plaque assay is a cell type that will fully support replication of the virus, typically A549 or HeLa cells. However, this procedure is lengthy, not only requiring incubation times of 7 or more days to form visible plaques, but also depending on good survival of the cell monolayer as well as efficient growth of the viral serotype. These criteria may not always be easy to achieve. Furthermore, replication-defective recombinant adenoviruses need to be assayed in a complementing cell line, such as 293, which often does not survive extended

From: *Methods in Molecular Medicine, Vol. 130:*
Adenovirus Methods and Protocols, Second Edition, vol. 1:
Adenoviruses Ad Vectors, Quantitation, and Animal Models
Edited by: W. S. M. Wold and A. E. Tollefson © Humana Press Inc., Totowa, NJ

periods in culture. This has led to the development of DNA-based or proteomic assays for recombinant adenoviruses *(3,4)*.

Physical or particle assays represent another approach to virus quantification *(5)* and are based on an empirically determined extinction coefficient for viral DNA *(6)*. These assays have the advantage of ease and simplicity but generate total viral particles per mL, rather than a measure of infectious units per mL. If the ratio of virus particles to infectious units varies between virus preparations, this can lead to problems in experimental or clinical study design *(5)*. Bioassays are thus the preferred assay of choice for quantitative determination of adenoviruses. The use of rapid bioassay methods, such as the fluorescent focus assay *(1,7)*, can overcome the problems associated with the conventional plaque assay. The fluorescent focus assay requires a short incubation time (approx 24 h) and labels fixed infected cells with a primary antibody (usually raised against the hexon protein) and a fluorescent secondary antibody. Fluorescent cells are counted by microscopy with UV optics, and a titer of fluorescent focus-forming units (FFU) per mL is obtained. To remove subjectivity in fluorescent cell counting and to analyze larger populations of infected cells, flow cytometric analysis has been employed *(8)*. Here we describe a simple procedure for flow cytometric assay of adenovirus type 5 (Ad5) using monoclonal antibodies against the viral hexon protein. This assay can also be used to identify and quantitate other adenovirus proteins in infected and transformed cells.

2. Materials

2.1. Target Cell Preparation and Viral Infection

1. Six-well tissue culture plates (Corning BV, The Netherlands).
2. Phosphate-buffered saline (PBS).
3. 1X Trypsin–ethylene diamine tetraacetic acid (EDTA) solution (Sigma, Poole, UK).
4. Serum-free Dulbecco's modified Eagle's medium (DMEM) (Sigma).

2.2. Target Cell Harvesting and Labeling for Flow Cytometry

1. 1X Trypsin-EDTA (Sigma).
2. PBS.
3. Paraformaldehyde (Sigma).
4. Triton X-100 (Sigma).
5. Normal goat serum (Vector Laboratories, Peterborough, UK).
6. Anti-hexon monoclonal antibody 88.1 *(9)* or 8C4 (Abcam, Cambridge, UK).
7. Anti-hexon antibody solution: 1 µL of anti-hexon antibody (1 µg) in PBS with 1% normal goat serum and 0.1% Triton X-100.
8. Isotype-matched control antibody. Dilute 1 µL of antibody (1 µg) in PBS with 1% normal goat serum and 0.1% Triton X-100.

9. Anti-mouse–FITC secondary antibody (Sigma).
10. Secondary antibody solution: dilute 2 μL of anti-mouse IgG, FITC conjugate (2 μg) in PBS with 1% normal goat serum and 0.1% Triton X-100.
11. A flow cytometer with a 488-nm laser and a suitable filter/detector (e.g., a Becton Dickinson FACSCalibur).

2.3. Target Cell Labeling for Microscopy

1. PBS.
2. Methanol (at –20ºC).
3. Anti-hexon monoclonal antibody (as in **Subheading 2.2.**, **item 6**).
4. Anti-mouse–FITC secondary antibody (Sigma) (or appropriate alternative).
5. Vectashield (Vector Laboratories).

2.4. Assessment of Focus Forming Units

1. Microscope with ×40 objective, UV optics, and appropriate filters.

3. Methods
3.1. Target Cell Preparation and Viral Infection

1. Plate target cells in 6-well dishes at a concentration such that they will be almost confluent the following day. Allow 14 wells of cells. If cells are to be assessed by microscopy, they should be seeded onto sterile cover slips (*see* **Note 1**).
2. After allowing the cells to adhere overnight, aspirate medium and wash each well twice with sterile PBS. Remove PBS from all cells.
3. Preprepare 10-fold dilutions of virus stock in serum-free medium, as shown in **Table 1**, and apply 200 μL to each well (duplicates for each dilution) (*see* **Note 2**).
4. Incubate at 37°C and 5% CO_2 for 1 h.
5. Add sufficient serum-free medium to each well to bring the total volume to 2 mL.
6. Incubate for a further 24 h.
7. Wash each well three times with PBS.
8. Proceed to microscopy or FACS analysis (*see* **Note 3**).

3.2. Target Cell Harvesting and Labeling for Flow Cytometry

1. Harvest infected cells using trypsin-EDTA, neutralizing trypsin with fresh medium containing serum, and count cells in a hemacytometer when detached.
2. Dispense an aliquot of 2×10^5 cells from each well into each of two 1.5-mL tubes. Prepare two tubes for each cell sample to be analyzed.
3. Spin cells at 300g for 5 min at room temperature and carefully remove supernatant.
4. Gently resuspend cells in 100 μL of 4% paraformaldehyde in PBS and incubate for 10 min at room temperature.
5. Centrifuge cells as above and remove supernatant. Resuspend pellet in 1% Triton X-100 in PBS. Incubate for 5 min at room temperature.
6. Centrifuge cells and resuspend the resulting cell pellet in 10% normal goat serum. Incubate at room temperature for 10 min.

Table 1
Preparation of Virus Dilutions for Assay by Either Microscopy or Flow Cytometry

Tube	Virus dilution	Serum-free medium	Virus
1	Control	400 μL	—
2	10^{-2}	495 μL	5 μL virus stock
3	10^{-3}	450 μL	50 μL of 10^{-2} dilution (tube 2)
4	10^{-4}	450 μL	50 μL of 10^{-3} dilution (tube 3)
5	10^{-5}	450 μL	50 μL of 10^{-4} dilution (tube 4)
6	10^{-6}	450 μL	50 μL of 10^{-5} dilution (tube 5)
7	10^{-7}	450 μL	50 μL of 10^{-6} dilution (tube 6)

7. Centrifuge cells again and resuspend one of each pair of cell pellets in 50 μL of anti-hexon antibody solution. Resuspend the second pellet in 1 μL of isotype-matched control (IMC) antibody. Incubate at room temperature for 1 h (*see* **Note 4**).

8. Centrifuge cells and resuspend pellets in 100 μL of PBS. Repeat wash step twice and remove supernatant.

9. Resuspend both cell pellets in 50 μL of secondary antibody solution. Incubate at room temperature for 1 h.

10. Wash cells once with PBS and resuspend in 300 μL of PBS. Transfer to FACS tubes.

3.3. Flow Cytometric Analysis

1. Set up FSC and SSC parameters, ideally using normal uninfected cells of the same type as those to be analyzed. Apply a gate (G1) to the cell population as shown in **Fig. 1**.

2. Set up FITC photomultiplier tube parameters using IMC antibody-labeled cells, gating on the G1 cell population. Ensure that control-labeled cells are in the center of the first decade, as shown in **Fig. 2**. Acquire 10,000 control-labeled G1-gated events and save data.

3. Without changing settings, acquire 10,000 G1-gated events of each anti-hexon-labeled cell sample.

3.4. Flow Cytometric Data Analysis

1. Display FITC histograms of control and anti-hexon-labeled samples. Set a marker on the uninfected hexon-labeled histogram, with less than 1% of cells being within the marker region, as shown in **Fig. 3**. Apply this marker region to the infected samples (**Fig. 4**). Obtain percentage of cells within the marker region.

2. To calculate FFU from the flow cytometric data will require calibration experiments for each individual system. In our system (*see* **Note 5**), the percentage of positive cells in the gated region was multiplied by 10^5 and then multiplied by the dilution factor to calculate FFU per 200 μL. This in turn is multiplied by 5 to give the FFU per mL (*see* **Note 5**).

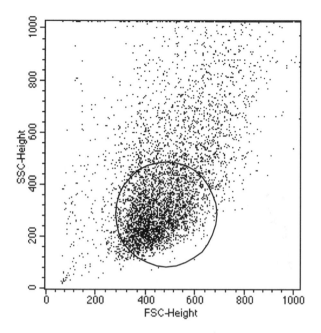

Fig. 1. FSC/SSC dot plot. Gate 1 (G1) is placed on live cell population and excludes cellular debris.

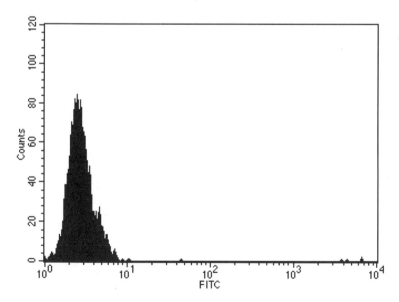

Fig. 2. Histogram of uninfected cells labeled with isotype matched control antibody. The photomultiplier tube settings should be adjusted so that the gated isotype-matched control peak is in the middle of the first decade.

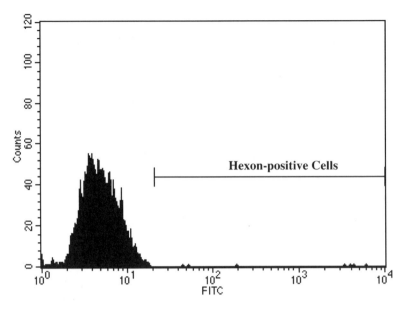

Fig. 3. Histogram of uninfected cells labeled with anti-Hexon antibody. The marker should be placed so that 2% or less of gated uninfected cells falls within the marker region. This 1% population represents nonspecific labeling.

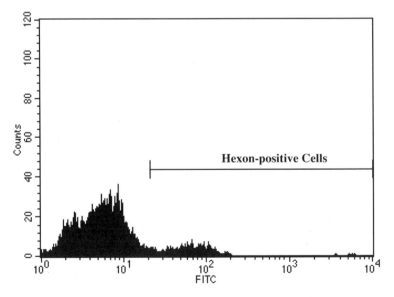

Fig. 4. Histogram of infected cells labeled with anti-Hexon antibody. The percentage of gated cells in the marker region (infected cells) should be determined from the statistics facility of the cytometry analysis program.

3.5. Target Cell Labeling for Microscopy

1. Wash cover slips twice with PBS at room temperature and aspirate.
2. Add 500 µL of ice-cold methanol (–20°C) to each cover slip and incubate for 5 min on ice.
3. Wash cover slips three times with PBS and aspirate.
4. Add 100 µL of anti-hexon antibody solution to each well: 1 µL of mouse anti-hexon antibody (1 µg) in PBS (plus 1% normal goat serum and 0.1% Triton X-100). Incubate at room temperature for 30 min.
5. Wash cover slips four times with PBS and aspirate.
6. Add 100 µL of secondary antibody solution: 2 µL of anti-mouse FITC (2 µg) in PBS (with 1% normal goat serum and 0.1% Triton X-100). Incubate at room temperature for 30 min.
7. Wash cover slips four times with PBS.
8. Mount cover slip on a slide using 1 drop of Vectashield.

3.6. Microscopical Assessment of FFUs

1. Examine at least 10 random fields of view using a ×40 objective (*see* **Note 6**).
2. Count fluorescent cells in each view. Average all counts from the 10 fields.
3. Multiply average count by 2500 to estimate positive cells per well.
4. Express cells/well as power (e.g., 1.5×10^5) and multiply by the dilution factor (e.g., 10^4). This gives FFU per 200 µL. Multiply this figure by 5 to give FFU per mL (*see* **Note 7**).

3.7. Concluding Remarks

Flow-cytometric analysis of adenovirus-infected cells provides a simple, objective method of virus bioassay. In contrast to counting of fluorescently labeled cells by microscopy, large numbers of cells can be rapidly analyzed and scored as hexon-positive using an instrument that is now available in most laboratories. In our hands, the bioactivity titers obtained by flow cytometry correlate well with those obtained by microscopy and fluorescent cell counting, but data are quicker to obtain and analyze. We have also used the flow cytometric protocol described above to detect and quantitate other intracellular adenovirus proteins such as E1A, providing an alternative to Western blotting.

4. Notes

1. It is important that the cells are approx 90% confluent at the point of infection. If the cells are not confluent, poor infection will result. However, overconfluence will also hinder efficient infection.
2. The infection range given here varies between 10^{-2} and 10^{-7} dilutions. However, the ideal range varies between virus preparations. If 10^{-2} gives insufficient infection, it may be necessary to use 10^{-1} or even, in the case of very low titer virus stocks, undiluted preparations.

3. This system shows good correlation between flow cytometry and microscopy. However, it will initially be necessary to perform parallel experiments with both detection methods to calibrate a particular system of virus, cell line, and antibody. Once calibration experiments have been performed, flow cytometry can be used alone to determine FFU.

4. The antibody used here is 88.1, which is an antibody directed against a group-specific determinant in the hexon protein and can therefore detect adenovirus serotypes from all of the subgroups. Although this antibody is not available freely, a similar antibody can be produced from a hybridoma cell line available from ATCC (2Hx-2 cell line, HB8177; A. Tollefson, personal communication). The assay described in this chapter also works well with 8C4, which is only one of several commercially available monoclonal antibodies against adenovirus hexon protein. Other antibodies of the same specificity may be substituted for 8C4 but will need to be calibrated and compared with the results obtained with microscopy.

5. For example, if 23% (or 0.23) of 10^4 gated cells are positive for hexon by flow cytometry at 10^{-3} dilution, the calculation will be $(0.23 \times 10,000 = 2.3 \times 10^3)$ 2.3×10^3 multiplied by the dilution factor 10^3, giving 2.3×10^6 FFU per 200 µL. This equates to 1.15×10^7 FFU/mL. It is important to calibrate the relationship between the traditional microscopy-based method and flow cytometry for individual systems. Different antibodies, target cell lines, or virus subtypes may affect the number of positive cells by flow cytometry.

6. When assessing FFU by microscopy, it is important to randomly select fields of view. The main source of error in this procedure is user subjectivity, such as only choosing fields of view with positive cells.

7. FFU sample calculation:
   Average positive cells per field (at 10^4 dilution) = 63.
   $63 \times 2500 = 157,500 = 1.6 \times 10^5$
   1.6×10^5 multiplied by 10^4 (dilution factor) = 1.6×10^9 = FFU per 200 µL
   1.6×10^9 multiplied by 5 = 8.0×10^9 FFU per mL.
   The multiplication factor of 2500 was derived from measurements of the field area with a graticule.

References

1. Precious, B. and Russell, W. C. (1985) Growth, purification and titration of adenoviruses, in *Virology: a Practical Approach* (Mahy, B. W .J., ed.), IRL Press, Oxford, pp. 193–205.

2. Tollefson, A. E., Hermiston, T. W., and Wold, W. S. M. (1999) Preparation and titration of CsCl-banded adenovirus stock, in *Adenovirus Methods and Protocols*. (Wold, W.S.M., ed.), *Methods in Molecular Medicine 21*, Humana Press, Totowa, NJ, pp. 1–9.

3. Kreppel, F., Biermann, V., Kochanek, S., and Schiedner G. A. (2002) DNA-based method to assay total and infectious particle contents and helper virus contamination in high-capacity adenoviral preparations. *Hum. Gene Ther.* **13,** 1151–1156.

4. Roitsch, C., Achstetter, T., Benchaibi, M., et al. (2001) Characterization and quality control of recombinant adenovirus vectors for gene therapy. *J. Chromatogr. B. Biomed. Sci. Appl.* **752,** 263–280.
5. Mittereder, N., March, K. L., and Trapnell, B. C. (1996) Evaluation of the concentration and bioactivity of adenovirus vectors for gene therapy. *J. Virol.* **70,** 7498–7509.
6. Maizel, J. V., White, D., and Scharff, M. D. (1968) The polypeptides of adenovirus. I. Evidence for multiple protein components in the virion and a comparison of types 2, 7a and 12. *Virology* **36,** 115–125.
7. Philipson, L. (1961) Adenovirus assay by the fluorescent cell-counting procedure. *Virology* **15,** 263–268.
8. Weaver, L. S. and Kadan, M. J. (2000) Evaluation of adenoviral vectors by flow cytometry. *Methods* **21,** 297–312.
9. Russell, W. C., Patel, G., Precious, B., Sharp, I., and Gardner, P. S. (1981) Monoclonal antibodies against adenovirus 5: preparation and preliminary characterization. *J. Gen. Virol.* **56,** 393–408.

17

Capture ELISA Quantitation of Mouse Adenovirus Type 1 in Infected Organs

Amanda R. Welton and Katherine R. Spindler

Summary

A capture enzyme-linked immunosorbent assay (ELISA) was optimized for identification of mouse adenovirus type 1 in infected brain homogenates. The ELISA method allows for a much faster quantitation of virus in infected organs than plaque assays. Methods for organ homogenization and the subsequent capture ELISA are described in this chapter.

Key Words: Mouse adenovirus; plaque assay; sandwich ELISA; capture ELISA; titration; homogenization; viral susceptibility; virus detection.

1. Introduction

Plaque assays are often used for quantitation of virus in cell and tissue samples, but they can be labor-intensive and lengthy. Mouse adenovirus type 1 (MAV-1) plaque assays take 10–12 d to develop and require counting on each of the last 3–4 d (*see* Chapter 4). We have developed a capture enzyme-linked immunosorbent assay (ELISA) method that requires only 5–6 h of "hands-on" time and one overnight step. We used regression analysis to compare the capture ELISA to the plaque assay and found that the capture ELISA method is a strong predictor of viral titer as determined by plaque assay. We therefore now use the capture ELISA method routinely for quantitation of MAV-1 from organs because it is a faster and more reproducible means of viral quantitation. This protocol could be adapted for use with other viruses for which antiviral antisera are available.

The first section of the chapter describes the preparation of infected mouse brain homogenates. The capture ELISA has been modified for spleen and liver

From: *Methods in Molecular Medicine, Vol. 130:*
Adenovirus Methods and Protocols, Second Edition, vol. 1:
Adenoviruses Ad Vectors, Quantitation, and Animal Models
Edited by: W. S. M. Wold and A. E. Tollefson © Humana Press Inc., Totowa, NJ

(K. Harwood, R. Rivera, and K. Spindler, unpublished) and could be optimized for other organs with little change to the overall protocol.

In the second section we describe the capture ELISA protocol. The protocol requires two antiviral antisera, which have been raised in two different species. For our assay we utilized polyclonal anti-MAV-1 *(1)* raised in rabbits against purified virions for the primary ("capture") antiserum. The IgG from the rabbit polyclonal serum was purified to increase the sensitivity of the assay. For the detecting antiserum, mouse polyclonal antisera were obtained from mice 12 wk after infection with MAV-1. The only criterion for the secondary antiserum is that it should be from a different species. For example, if the primary antiserum is from rabbit, then the secondary can be from mouse, rat, goat, or other lab animals.

In the third section we describe the steps used to optimize a capture ELISA for a different virus. Optimization of the antisera, incubation, wash, and detection conditions are discussed together with the appropriate controls *(2)*.

2. Materials

1. Ethanol.
2. Sterile scalpel handle and disposable sterile blades.
3. Biospec BeadBeater (BioSpec Products, Bartlesville, OK, cat. no. 1001) (*see* **Note 1**).
4. 1.0-mm Glass beads (BioSpec Products, Bartlesville, OK, cat. no. 11079110).
5. 2-mL Screw-cap vials (BioSpec Products, Bartlesville, OK, cat. no. 10832) (*see* **Note 2**).
6. 1.5-mL Microfuge tubes.
7. Primary antiviral IgG purified from rabbit anti-MAV-1 antiserum (AKO1-103) *(1)* by DEAE Affi-Gel Blue chromatography (Bio-Rad Laboratories, Hercules, CA).
8. Eppendorf 12-channel pipet (Fisher Scientific, Pittsburgh, PA, cat. no. 2137895).
9. 200-µL Pipet tips.
10. 0.1 *M* Sodium carbonate (Na_2CO_3) buffer, pH 9.6.
11. Immulon 2 HB 96-well plates, Thermo Labsystems, Franklin, MA (Fisher Scientific, Pittsburgh, PA, cat. no. 1224561).
12. Dispensing basins (Fisher Scientific, Pittsburgh, PA, cat. no. 13681100).
13. Igepal (Sigma-Aldrich, St. Louis, MO, cat. no. I-3021) (NP40 can be substituted for Igepal).
14. 1.5 *M* NaCl.
15. 1 *M* Tris-HCl, pH 8.0.
16. Lysis buffer: 0.15 *M* NaCl, 1% Igepal, 50 m*M* Tris-HCl, pH 8.0 (*see* **Note 3**).
17. 1-L Corning glass bottles.
18. Plastic squirt bottle.
19. Nutator (Clay Adams, Parsippany, NJ).
20. 1% Bovine serum albumin (BSA) (Sigma-Aldrich, St. Louis, MO, cat. no. A-7906) in water (*see* **Note 4**). Filter-sterilize and store at 4°C.

21. Eppendorf 20- to 300-μL tips (Fisher Scientific, Pittsburgh, PA, cat. no. 0540342).
22. 200-μL Plugged pipet tips.
23. 1X Phosphate-buffered saline (PBS). Autoclave and store at 4°C.
24. Mouse brain sample (antigen).
25. MAV-1 virus stock (positive control).
26. Nalgene dishpan (Fisher Scientific, Pittsburgh, PA, cat. no. 1335920A).
27. 7-mL Vials (Sarstedt, Newton, NC, cat. no. 60.542).
28. Secondary antiviral serum: mouse anti-MAV-1, heat-inactivated for 1 h at 57°C.
29. Anti-mouse IgG horse radishperoxidase (HRP)-conjugated (Amersham, Piscataway, NJ, cat. no. NA931).
30. 1 *M* Tris-HCl, pH 7.2.
31. 1 *M* NaCl.
32. Bovine serum, heat-inactivated for 1 h at 57°C.
33. Conjugate dilution buffer: 0.01 *M* Tris-HCl, pH 7.2, 0.15 *M* NaCl, 15% bovine serum (Invitrogen, Carlsbad, CA) (*see* **Note 5**). Filter-sterilize and store at 4°C.
34. 1-Step Turbo TMB-ELISA (Pierce, Rockford, IL, cat. no. 34022).
35. 0.9 *M* H_2SO_4.
36. Plate reader/spectrophotometer (capable of reading OD_{450} in 96-well format).

3. Methods

3.1. Mouse Brain Homogenization

1. For each sample to be homogenized, label an autoclaved 2-mL screw-cap vial that is three-quarters filled with 1.0-mm glass beads (*see* **Note 2**). In addition, label two 1.5-mL microfuge tubes for each sample (*see* **Note 6**).
2. Tare the labeled 2-mL screw-cap vials and record the weight on the side of vial using an ethanol-resistant marker.
3. Attach a sterile scalpel blade to a sterile scalpel handle, dip in ethanol, and flame to burn the ethanol off.
4. In a biosafety cabinet, cut a small piece of virus-infected or mock-infected brain tissue (approximately a cube, 4 mm on a side), place in 2-mL screw-cap vial, and weigh. Subtract the tube weight from the total to get the weight of the brain sample and record. Use roughly 100 mg of brain per sample, no less than 90 mg (*see* **Note 7**). Keep on ice.
5. Continue for all samples, flaming the scalpel in between. Change the scalpel blade between mock-infected and virus-infected samples and also between samples from different experimental conditions (mouse strain, dose, virus strain, etc.).
6. When all samples have been cut and weighed, add 1 mL cold 1X PBS to each tube containing beads and brain sample (*see* **Note 8**). Keep on ice.
7. Homogenize using BeadBeater as per manufacturer's instructions for a total of 3 min in 1-min intervals with 1-min rests between.
8. Remove homogenate to first microfuge tube using a plugged micropipet tip (*see* **Note 9**). Plunge pipet tip into the beads to remove all homogenate. Beads should

not be aspirated into the tip when using either a P200 or P1000 size. Continue for all samples, changing tips for each sample.

9. Spin homogenates at 700*g* for 5 min at room temperature.
10. Remove supernatant to second microfuge tube and store at –20°C (*see* **Note 10**).
11. Thaw and freeze homogenates at least three times to ensure release of virus from the cells.
12. Dilute all homogenates to 9% w/v in 160 µL 1X PBS (*see* **Note 11**). Store at –20°C.

3.2. Antigen-Capture ELISA

1. Prepare a 1:100 stock of capture antiserum, unpurified or DEAE Affi-Gel Blue Gel-purified, in 1X PBS. Store at –20°C. From this stock prepare a 1:2000 dilution in 0.1 *M* sodium carbonate buffer for the number of wells needed (*see* **Note 12**) and keep on ice.
2. Pour the prepared 1:2000 dilution of capture antiserum into a dispensing basin and set multichannel pipettor to 50 µL.
3. Dispense 50 µL of diluted capture antiserum into each well of an Immulon 2 HB 96-well plate to coat each well (*see* **Notes 13–15**).
4. Use a second 96-well plate as a cover and incubate overnight at 4°C on a Nutator (*see* **Note 16**).
5. On the following day, dump the well contents into a sink and manually wash the plate (*see* **Note 17**) four times with lysis buffer (**Subheading 2.**, **item 16**).
6. Block wells by dispensing 300 µL of 1% BSA into each well and incubate for 1 h at room temperature on a Nutator (*see* **Note 18**).
7. Dump, wash plates, and "bang dry" as before (**step 5**).
8. In a biosafety cabinet, dispense 50 µL of 9% w/v mouse brain homogenate (*see* **Subheading 3.1.**) or required controls to the appropriate wells (*see* **Notes 13** and **19**). Incubate for 1 h at room temperature on a Nutator.
9. In a biosafety cabinet, dump the contents of the wells into a Nalgene dishpan (*see* **Note 20**). Wash plates and "bang dry" as previously.
10. Prepare a 1:250 dilution of secondary ("detecting") antiserum (mouse anti-MAV) in 1% BSA (*see* **Subheading 2.**, **item 20**). Dispense 50 µL per well and incubate for 1 h at room temperature on a Nutator.
11. Remove well contents, wash plates, and dry as previously (**step 9**).
12. Prepare a 1:100 dilution of HRP-conjugated anti-mouse IgG in conjugate dilution buffer (**Subheading 2.**, **item 33**) and store at 4°C for up to 2 wk. On the day of the assay, prepare a 1:5000 dilution in conjugate dilution buffer from the 1:100 stock and dispense 50 µL per well. Incubate for 1 h at room temperature.
13. During this incubation, warm the required amount of Turbo-TMB substrate to room temperature (*see* **Note 21**).
14. Dump well contents, wash plates five times, and dry thoroughly (*see* **Note 22**).
15. Dispense 125 µL of Turbo-TMB into each well. Allow the plates to develop at room temperature for approx 15 min or until the negative control begins to turn blue (*see* **Notes 23** and **31**).

16. To stop the reaction, add 125 µL of 0.9 *M* sulfuric acid to each well (*see* **Note 24**).
17. Immediately read the OD$_{450}$ on a plate reader/spectrophotometer.
18. Subtract the mean of the mock-infected homogenate values (background) from the experimental sample values.

3.3. Antigen-Capture ELISA Optimization

1. Optimize dilution of primary antiviral serum. An initial step will be the determination of the appropriate time for collection of antiserum from infected animals (*see* **Note 25**). Optimization will start with a range of dilutions from 1:500 through 1:2000. This step may be further optimized by changing the buffer used to dilute the antiserum. The 0.1 *M* sodium carbonate (Na$_2$CO$_3$) buffer listed above has worked best under the conditions described here, but a borate buffer could be used as well (*see* **Note 26**).
2. Optimize blocking time and temperature. In addition to 1 h at room temperature, try 1 h at 37°C. Increasing the time at either temperature may also reduce background.
3. Optimize homogenate preparation by assaying undiluted and various dilutions of homogenate (*see* **Note 27**). Try 1:10 through 1:100 dilutions of homogenate in 1X PBS.
4. Optimize secondary antiviral antiserum (*see* **Note 28**). An initial step will be to determine the appropriate time point of harvest whether this antiserum is from infected animals or immunization with purified virus (*see* **Note 29**). Assay using various serum dilutions. Start with a range of 1:200 through 1:1000.
5. Optimize tertiary antiserum (*see* **Note 30**). This antiserum must be an anti-IgG HRP-conjugated antibody (*see* **Note 23**). Test various dilutions of this antibody. Start with a range of 1:1000 through 1:10,000.
6. When all steps have been optimized for the best signal, test the ELISA system by successively leaving out antibodies and using undiluted virus as antigen. For instance, to test whether the primary (coating) antiviral antiserum is specific, replace it in one assay with 1X PBS and run the rest of the steps as normal. If the primary antiserum is specific and the other steps show little nonspecific binding, then this step should produce no signal above background. To test the secondary and tertiary antisera, repeat the above steps, substituting 1X PBS for each and assaying as usual. Once all steps have been optimized, these controls need not be run in each assay.

4. Notes

1. If only a few samples are to be analyzed at a time, a Mini BeadBeater can be purchased from BioSpec Products (cat. no. 3110BX).
2. Prior to starting the homogenization, 2-mL screw-cap vials should be three-quarters filled with 1.0-mm glass beads and autoclaved.
3. Prepare the lysis buffer in 1-L glass bottles. After adding components of the lysis buffer, mix well, and place at 37°C until Igepal has fully dissolved. Test by swirling bottles and noting if any residual Igepal comes up off the bottom of the bottles. Store at 4°C until use.

4. When preparing 1% BSA, stir no longer than 10 min.

5. The bovine serum used in this buffer does not need to be of high quality.

6. The screw-cap vial will be used for the homogenizing. The first microfuge tube will contain the homogenate to be spun down, and the final microfuge tube will contain the completed homogenate.

7. All homogenates will be adjusted to 9% w/v for the capture ELISA. This percentage will not be possible if samples weigh less than 90 mg. Homogenates can be prepared from fresh or frozen tissue.

8. It may be necessary to dispense liquid directly into beads by inserting the pipet tip into the dry beads to get the entire 1 mL into the tube. It is important to add exactly 1 mL to each tube for calculation of concentration. Be sure that the cap and the washer surrounding the cap interior are free of beads, otherwise there may be leakage during homogenization.

9. Use plugged tips for all steps involving homogenate, which could contain infectious virus.

10. For each sample you will recover slightly less than 1 mL of homogenate.

11. Each sample homogenate should be assayed in triplicate (3 wells, 50 µL each) to allow for appropriate statistical analysis of the assay. Therefore, prepare 160 µL of homogenate diluted to 9% to allow for error in pipetting. For example, if the weight of the sample determined in **step 4** is 0.1390 g and you added 1 mL of 1X PBS, the w/v is 13.90%. To adjust the sample to 9% w/v, calculate as follows: (9/13.90) × (0.160) to have 160 µL of 9% w/v homogenate. Repeat for all samples.

12. Prior to preparing the capture antibody, label a 96-well paper grid with the samples and controls to be run. Use this to calculate the amount of each antibody needed. For example, if 96 wells are being used, a total of (96 wells) × (0.05 mL per well) = 4.8 mL of diluted antibody is needed. Always make extra to allow for pipetting error.

13. In addition to assaying each experimental sample in triplicate, include triplicate wells of 1X PBS, mock homogenate, and conditioned medium (as control for virus stock) as negative controls. Also include triplicate wells of undiluted virus as positive control.

14. For all dispensing steps, angle the tips against the side of the well when dispensing liquid to lessen the risk of detaching antibodies or complexes by force of dispensing.

15. After dispensing coating buffer, tilt plate on its edge and gently tap to ensure the liquid has covered the well. Do not push liquid around with the micropipet tip, as this will remove coating buffer from the wells.

16. Use a second 96-well plate as a cover throughout all incubation steps. Use a large rubber band to hold plates on the Nutator during rocking.

17. Fill a plastic squirt bottle with lysis buffer. After dumping the well contents, squirt wells full of buffer by squirting at the sides of the wells while holding the plate at an angle, so as not to dislodge antibodies from well bottom. Move quickly from well to well. When washing is completed, "pound" or "bang" plate upside

down on a stack of paper towels until all liquid is gone and few bubbles remain in the wells. This should be done quite forcefully.

18. A volume of 300 µL will fill the wells. This is required to ensure complete blocking for subsequent steps.

19. The brain homogenates and virus controls are infectious.

20. The wells at this step will contain infectious virus. Bleach treatment of the discarded well contents in a dishpan is required. In addition, bleach the sink following washes.

21. 125 µL of the Turbo-TMB is needed per well. Warm only the amount to be used, not the entire bottle, which should be stored at 4°C.

22. For the final washing step, be sure to do an extra wash (for a total of five washes) and bang the plate out thoroughly.

23. Development usually takes 10–15 min. During that time monitor the blue color change in the negative control wells that have mock-infected homogenate. If these wells begin to show color change, stop the reaction with the sulfuric acid addition. The 1-Step Turbo TMB used in **step 15** is an HRP substrate. When oxidized with hydrogen peroxide (catalyzed by HRP), the TMB yields a blue color.

24. The addition of sulfuric acid will change the color in the wells from blue to yellow. This color will continue to darken over time, changing your reading. Therefore, it is important to get a reading of the OD as soon as possible.

25. We use antiviral antiserum raised in rabbits as primary (coating) antibody. This antiserum was obtained 103 d postimmunization with purified MAV-1 *(1)*. It may be necessary to test antibody titers from various bleeds to ensure optimal specificity.

26. Borate buffer, pH 8.5: Make 1 L containing 6.19 g boric acid, 9.5 g sodium borate ($Na_2B_4O_7 \cdot 10\ H_2O$), and 4.39 g NaCl. Adjust pH to 8.5 with NaOH or HCl. Filter-sterilize and store at 4°C.

27. The tissue being used (if other than brain) may interfere with antibody interactions. In that case, it may be necessary to dilute the homogenate in order to decrease the amount of debris. Start with only a 1:10 dilution for the antigen step. If a dilution is used, dilute all samples the same amount in subsequent assays.

28. It is crucial that the secondary antiviral antiserum is raised in a different species than the primary antiviral antiserum.

29. As with the primary antiviral antiserum, it may be necessary to test antibody titers from various bleeds to ensure optimal specificity.

30. The tertiary antiserum must be raised against the animal used to make the secondary antiserum.

References

1. Kajon, A. E., Brown, C. C., and Spindler, K. R. (1998) Distribution of mouse adenovirus type 1 in intraperitoneally and intranasally infected adult outbred mice. *J. Virol.* **72,** 1219–1223.

2. Crowther, J. R. (ed.) (1995) *ELISA: Theory and Practice.* Humana Press, Totowa, NJ.

18

Preparation and Titration of CsCl-Banded Adenovirus Stocks

Ann E. Tollefson, Mohan Kuppuswamy, Elena V. Shashkova, Konstantin Doronin, and William S. M. Wold

Summary

Adenovirus research often requires purified high-titer virus stocks and accurate virus titers for use in experiments. Accurate titers are important for quantitative, interpretable, and reproducible results. This is especially true when there are comparisons of different mutant viruses following infection. This chapter details the large-scale preparation of adenovirus (either replication-competent or replication-defective) in spinner cultures (e.g., KB, HeLa, or 293 cells). Protocols for harvesting cells and isolation of adenovirus by CsCl banding are presented. Methods for titering adenovirus by plaque assay are presented along with a discussion of how plaque assays can be used to determine the kinetics of cell killing and cytolysis by adenoviruses.

Key Words: Adenovirus; CsCl-banded; titration; plaque assay; adenovirus death protein; plaque development; A549; 293; suspension culture; large-scale virus preparation.

1. Introduction

An important step in the development of modern experimental virology was the development of the plaque assay: first with bacteriophage, then with eukaryotic viruses. In order to obtain quantitative, interpretable, and reproducible results, it is necessary to know how much virus is being used in the experiment. With adenoviruses, several approaches have generally been used to quantitate virus stocks. First, virus particles are counted, e.g., in an electron microscope *(1,2)*. The problem with this approach is that many adenovirus particles are not infectious, perhaps because they have a defective complete genome or they lack fiber or some other protein. The second approach is to determine the number of plaque-forming units (PFU) per mL. Here, the analy-

From: *Methods in Molecular Medicine, Vol. 130:*
Adenovirus Methods and Protocols, Second Edition, vol. 1:
Adenoviruses Ad Vectors, Quantitation, and Animal Models
Edited by: W. S. M. Wold and A. E. Tollefson © Humana Press Inc., Totowa, NJ

sis quantitates the number of virions capable of a full infectious cycle. This approach will be described in detail in this chapter.

Experimental reproducibility also requires that adenovirus stocks be prepared in a consistent manner. Such stocks are stable for years when stored at –70°C. One simple approach is to prepare a cytopathic effect (CPE) stock, in which an isolated plaque is picked and a small dish of permissive cells (e.g., A549 or 293) is infected. After about 4–5 d, the cells in the monolayer will show typical "adenovirus CPE", i.e, their nuclei will become enlarged and they will round up and detach from the dishes into individual floating cells as well as grape-like clusters. These floating cells remain alive for some time (cell death and the release of adenovirus from the cells begins approx 3 d postinfection for subgroup C adenoviruses). Cells are collected, adenovirus is released by repeatedly freezing and thawing, and then the lysate is used to infect a larger monolayer, such as a T-150 flask. After CPE appears, cells are collected and adenovirus is released by three rounds of freeze–thawing followed by sonication, and the adenovirus is titered by plaque assay.

These CPE stocks, which are typically 10^8–10^{10} PFU/mL, are adequate for exploratory studies. Large-scale virus stocks are usually prepared by banding the virus in CsCl equilibrium density gradients, and this procedure will be described. CsCl banding yields large quantities of high-titer (10^{11} PFU/mL) adenovirus stocks.

2. Materials

2.1. Cell Culture Media and Stock Solutions

1. Dulbecco's modified Eagle's medium (DMEM) (with high glucose, with L-glutamine, with phenol red, without sodium pyruvate, without sodium bicarbonate; Invitrogen, or JRH Biosciences).
2. Sodium bicarbonate (tissue culture grade; Invitrogen or Sigma).
3. Penicillin/streptomycin stock (1000X): 10,000 U/mL of penicillin G sodium and 10,000 µg/mL streptomycin sulfate in 0.85% saline (Invitrogen). Store at –20°C.
4. 0.22-µm Filters (Millipore, Corning, or Nalge).
5. Minimum essential medium (S-MEM) (Joklik-modified) (with L-glutamine, with 10X phosphate, without sodium bicarbonate; Invitrogen or JRH Biosciences).
6. DMEM (20 L): Dissolve DMEM (2 × 10 L) completely in 19 L of H_2O (tissue culture grade), then add 74 g (3.7 g/L) of sodium bicarbonate. Adjust the pH of the solution to approx 6.9 with 1 N HCl or 1 N NaOH (pH will increase 0.3 × 0.4 units with filtration). Add penicillin-streptomycin stock (1 mL/L; Invitrogen). Medium is membrane-sterilized by positive or negative pressure (positive pressure is preferable in maintaining pH) through a 0.22-µm filter (*see* **Note 1**).
7. S-MEM (Joklik-modified) for Suspension Cultures (20 L): S-MEM (Joklik-modified) (2 × 10 L), 40 g sodium bicarbonate (2 g/L). pH is adjusted to 6.9 with 1 N HCl or 1 N NaOH; then streptomycin–penicillin stock is added (1 mL/L). Sterilize by membrane filtration (0.22-µm filter).

8. 2X DME for plaque assay overlays (5 L): combine 4.5 L ddH$_2$O (tissue culture grade), 0.6 g penicillin G (sodium salt, 1670 U/mg), 1.0 g streptomycin sulfate (787 U/mg), DMEM powder (1 × 10 L). Adjust to 5 L; pH is not adjusted at the time of preparation (sodium bicarbonate stock is added at the time of overlay preparation). Sterilize by membrane filtration through a 0.22-μm filter. 2X DME is aliquoted in 500-mL volumes and is stored at 4°C.

9. Dulbecco's phosphate-buffered saline (PBS) (without calcium chloride, without magnesium chloride; Sigma). Prepare with tissue culture grade water and sterilize by filtration through a 0.22-μm filter.

10. Trypsin (1:250; Difco Laboratories).

11. Phenol red (sodium salt; Fisher).

12. Penicillin G (sodium salt; 1670 U/mg, Sigma).

13. Streptomycin sulfate (787 U/mg; Sigma).

14. Trypsin–ethylene diamine tetraacetic acid (EDTA) stock solution (4 L): dissolve 4 g trypsin, 2 g EDTA (disodium salt), 4 g dextrose, 20 mg phenol red, 0.25 g penicillin G, 0.45 g streptomycin sulfate in a final volume of 4 L of Dulbecco's PBS (calcium- and magnesium-free); adjust pH to 7.2. Sterilize by membrane-filtration. Aliquoted stocks are stored frozen until the time that they are needed as working stocks.

15. Horse serum (HS) and fetal bovine serum (FBS): Sera are needed for supplementation of the tissue culture media (*see* **Notes 2** and **3**). Horse serum is normally heated (56°C for 30 min) to inactivate complement prior to use in growth or during infection of KB cells.

16. Ultra-pure glycerol (Invitrogen).

17. Tris-saline-glycerol (TSG) (for dilution of cesium chloride-banded virus): preparation of solutions A and B:
    a. Solution A: 900 mL ddH$_2$O, 8.0 g Na$_2$Cl, 0.1 g NaHPO$_4$ (dibasic), 0.3 g KCl, 3.0 g Trisma base (Tris); adjust pH to 7.4 by addition of approx 1–2 mL of concentrated HCl. Adjust volume to 1 L.
    b. Solution B: 2.0 g MgCl$_2$, 2.0 g CaCl$_2$, 100 mL ddH$_2$O. Combine 700 mL of solution A with 3.5 mL of solution B; add 300 mL of ultra-pure glycerol. Heat solution in microwave and filter-sterilize through a 0.22-μm filter (unheated solution is too viscous to filter).

18. Dialysis buffer (5% sucrose buffer): 25 mM Tris, pH 7.4, 140 mM NaCl, 5 mM KCl, 0.6 mM Na$_2$HPO$_4$, 0.5 mM MgCl$_2$, (0.9 mM CaCl$_2$), (5% w/v sucrose). Prepare 2 L of a 10X solution (autoclave) lacking the CaCl$_2$ and sucrose (CaCl$_2$ will cause precipitation of salts in the 10X stock and sucrose should not be autoclaved). To prepare 2 L of a 1X sucrose buffer, mix: 200 mL 10X stock solution with 1.76 L ddH$_2$O, and add 100 g sucrose and 1.8 mL 1 M CaCl$_2$ solution.

19. Difco Agar Noble (1.8% Agar Noble stock): Add 1.8 g of Noble agar per 100 mL of tissue culture-grade H$_2$O (9 g of Noble agar in 500 mL of water is convenient). Autoclave for 30 min to sterilize. Store at 4°C.

20. 7.5% (w/v) Sodium bicarbonate stock solution (Invitrogen).

21. Glutamine stock (200 mM; Invitrogen).

22. Neutral red stock (3.333 g/L of neutral red sodium salt in Dulbecco's PBS, membrane-filtered; Sigma).

23. Agar overlay medium for plaque assays. Final volumes of components for 100 mL of overlay are as follows: 50 mL of 2X DME, 5 mL of 7.5% (w/v) sodium bicarbonate stock solution, 2 mL of FBS, 43 mL of 1.8% agar Noble stock. Microwave 1.8% agar Noble stock to melt, then reduce temperature to 56°C in a 56°C water bath prior to addition to the other overlay components. Mix the other components of the overlay, and keep at 37°C in a water bath. These temperatures are used to keep the agar from solidifying while also ensuring that the overlay will not be hot enough to result in cell killing when it is layered onto the monolayer. If stock of 2X DME was prepared 1 mo or more prior to the date of plaque assays, add 1 mL of glutamine stock (200 mM) per 100 mL of overlay. Neutral red is added to the second overlay (*see* **Note 4**); add 0.45 mL of neutral red stock (3.333 g/L) per 100 mL of overlay.

24. Globe-shaped Florence boiling flasks (long neck, flat-bottom): gradation of sizes from 250 mL to 6 L (Pyrex or Kimble) for expansion of spinner culture cells (*see* **Note 5**).

25. Silicone stoppers for flasks (Fisher Scientific or other suppliers): sizes 5, 6, 8, 10, 11 (or 29D, 31D, 39D, 41D, 49D, 60D [European sizing]).

26. Dumbbell-style stir bars: 32 mm and 52 mm (Fisher Scientific).

27. Crystal violet staining solution: 1% crystal violet, 20% ethanol, 10% formaldehyde in water.

28. Slide-A-Lyzer dialysis cassettes (10,000 MWCO; Pierce).

3. Method

3.1. Growth of Monolayer and Suspension Cultures

3.1.1. KB Cells (see **Note 6**)

KB suspension cultures are a clonal line derived in the laboratory of Maurice Green (from a KB suspension culture received originally from Dr. Harry Eagle) and grown in the laboratories of Maurice Green and William Wold. This clonal line is reported to produce higher yields of virus than the parental KB cell line *(3,4)*.

1. KB cells are grown in suspension in Joklik-modified minimum essential medium (MEM) (5% heat-inactivated horse serum) in Florence boiling flasks (Pyrex or Kimax; *see* **Note 5**). Cells are maintained in culture with daily dilution of cells to maintain cultures in the 1.5–4.0 × 10^5 cells/mL range.

2. Split the culture each day to a cell density of 1.5–2.0 × 10^5 cells per mL. Cells typically double in 24 h; a volume of new medium is added to the cell suspension, with excess cells being discarded. Only a portion of the medium is replaced because "conditioned" medium appears to have some beneficial effect on the growth of the cells (perhaps from autocrine effects).

3.1.2. 293 Spinner Cultures

It is possible to do large-scale Ad vector preparations in 293 cells (ATCC or Microbix) that have been put into suspension culture in 5- to 6-L spinner flasks. This is both a great cost- and time-saver. Suspension culture cells can be grown in medium containing horse serum (5% v/v) rather than calf or fetal calf serum. Use the following procedure to adapt 293 monolayers to spinner culture (*see also* **Note 7**):

1. 293 cells which are nearly confluent on 150-mm tissue culture dishes are washed two times with Joklik's medium (serum-free).
2. Add Joklik's medium containing 5% HS and return to 37°C and 6% CO_2 in an incubator from overnight to 2 d.
3. Cells will have rounded up to some degree. Using the medium on the dish, pipet medium across the 293 cells to remove the cells from the dish. Count cells on a hemacytometer.
4. Dilute cells to 4×10^5 per mL and transfer into a 100-mL spinner flask (volume should be at least 40–50 mL to maintain appropriate pH).
5. On the next day count the cells; cells should be between 3×10^5 and 4×10^5 cells/mL. Subsequently, split the 293 cells to 1.5×10^5 cells per mL every 1–2 d (diluted by addition of fresh medium; do not remove previous "conditioned" medium). Do not allow cells to reach a density higher than 4×10^5 cells per mL.
6. Expand cells to the required volume and cell number for infection (by moving through a series of spinner flasks of increasing volumes). This will normally take about 7–10 d to reach a volume of 6 L. Follow the steps in **Subheading 3.2.** for infection of spinners.

3.1.3. A549 Cells

A549 cells (CCL 185; American Type Culture Collection, Rockville, MD) are grown in DMEM supplemented with glutamine and 10% FBS (Hyclone, BioWhittaker, or Invitrogen).

1. For routine passage, medium is removed from the plates and trypsin-EDTA is added (2 mL for 100-mm dish or 5 mL for a 175-cm² flask).
2. When cells have rounded up (usually in 3–5 min), cells are removed from the dish by adding DMEM (10% FBS) with gentle pipetting. Use of 10% FBS/DMEM inhibits continued trypsin action, and cells display higher viability.
3. Cells are centrifuged at 100–250*g* in a tabletop centrifuge (e.g., Beckman GS-6) to pellet cells.
4. Trypsin/medium solution is removed, and cells are resuspended in DMEM (10% FBS) and plated at 1:5 to 1:20 dilutions relative to the original cell density.
5. Cells are usually passaged at 2- to 3-d intervals. Cells are grown at 37°C with 6% CO_2 in humidified incubators in 100-mm dishes or 175-cm² flasks. Cells should not be allowed to become very heavy in routine culture or they will not have good survival in plaque assays.

3.2. Large-Scale Adenovirus Preparation

Spinner KB or 293 cells are used for large-scale production of adenoviruses. Cells are grown in MEM, Joklik-modified (a suspension medium with reduced calcium and increased levels of phosphate) with 5% HS (heat-inactivated at 56°C for 30 min to inactivate complement). For infection, it is typical to use a 3- to 6-L volume of cells that have reached a density of 3–3.5×10^5 cells/mL. For a 3-L infection, the volume is reduced to 1 L by centrifuging 2400 mL of the suspension culture and then resuspending the pelleted cells in approx 400 mL of Joklik-modified MEM (*see* **Note 8**) and returning cells to the spinner. Cells are infected with 5–20 PFU/cell with stock viruses (lower multiplicity of infection [MOI] will result in fewer defective particles). Virus is adsorbed with spinning at 37°C for 1 h; at the end of the adsorption period, 2 L of medium (with 5% HS) are added to the infection. Infected cells are maintained in spinner flasks at 37°C for 40–46 h before harvesting. Given viruses may be incubated for more extended times if cell lysis is not occurring (if the E3 ADP gene is absent or mutated or when working with Ad vectors in 293 cells).

Day 1

1. Prepare 3 L (or 6 L) of spinner cells at approx 2×10^5 cells per mL in Joklik-modified MEM/5% HS.

Day 2

1. Use a hemacytometer to do a cell count to determine the cell number. This cell number will be used to determine the volume of virus to use for infection.
2. For infection of 3 L of cells, reduce the total cell volume to 1 L by centrifugation. Cells (2400 mL) are pelleted in a tabletop centrifuge (Beckman GS-6 centrifuge) in 750-mL Beckman centrifuge bottles at $250g$ for 10 min. Do not brake at the end of the spin (the centrifuge bottle bottoms are flat and cells will not adhere to the bottom if brake is applied). For 293 cells, remove much of the medium over the pellet by aspiration of the supernatant (293 cell pellets will be very loose). For KB cells, the supernatant can be carefully poured from the centrifuge bottles (leaving the last 100 mL of medium over the pellet).
3. Cells are resuspended in approx 400 mL of Joklik-modified medium (~100 mL per centrifuge bottle) and returned to the spinner flask.
4. Add virus (5–20 PFU/cell or use a portion of a CPE stock from a tissue culture dish or flask). If using small volumes of banded virus, it is best to dilute the virus in serum-free Joklik-modified MEM in a 50-mL centrifuge tube (Falcon, Corning) prior to addition to the spinner. Adsorb for 1 h at 37°C with spinning. Then add Joklik's medium containing 5% HS to return the total volume to the starting volume of 3 L (or 6 L).

Day 4

1. Place a portion of the cell suspension on a hemacytometer to evaluate the infection (*see* **Note 9**). Infections of KB spinners are typically harvested 40–46 h postinfection. With viruses that have E1A mutations or for Ad vectors (E1-deleted) that are being grown in 293 spinner cultures, it will likely be necessary to harvest at 60–72 h postinfection.
2. Pellet infected cells in 750-mL Beckman centrifuge bottles in tabletop centrifuge (250g for 10 min at 4°C). Do not use brake at end of centrifugation.
3. Remove medium; resuspend cell pellets in a total volume of 150–200 mL of cold PBS (4°C) and transfer to a 250-mL conical centrifuge tube (Corning). Centrifuge at 250g at 4°C for 10 min.
4. Repeat PBS wash and pelleting of cells two times.
5. Resuspend cell pellet in enough sterile, cold (4°C) 10 mM Tris-HCl, pH 8.0, to give a final volume of 24 mL (*see* **Note 10** for an alternate extraction protocol).
6. Aliquot 8 mL into each of three sterile polypropylene snap-cap tubes (15-mL size), wrap caps with Parafilm, and freeze at –70°C or in ethanol/dry ice bath for at least 1 h (processing can be left at this point for one or more d before completing the remainder of the protocol).
7. Thaw tubes in 37°C water bath. Repeat these freeze–thaw steps two additional times and then place tubes on ice.
8. Cells are disrupted by sonication on ice in the cup of a Branson Sonifier 250. Settings are as follows: duty cycle on "constant," output control on 9 (scale of 1–10), and 3-min cycles. This is repeated three times for each sample.
9. Transfer sonicated material to a sterile 50-mL flip-cap centrifuge tube and centrifuge at 12,000g for 10 min at 4°C in a Beckman J2-HC centrifuge; then remove supernatant (which will contain released virions) and discard cell debris pellet.
10. Determine the volume of supernatant and multiply by 0.51. The resulting number will be the grams of CsCl to be added to the preparation; for example: 20 mL of supernatant × 0.51 g of CsCl per mL = 10.2 g of CsCl for addition to the supernatant.
11. After mixing with CsCl, divide sample into two Ti50 quick-seal tubes.
12. Centrifuge in Ti50 rotor at 110,000g in Beckman ultracentrifuge at 4°C for 16–20 h to band the virus.

Day 5

1. Stop the ultracentrifuge without using the brake.
2. The virus will appear as a white band, which will be approximately at the middle of the tube. Collect by syringe-puncture at the bottom (puncture top of tube as well). Band will visibly move down the tube; collect the band region in a 15- or 50-mL sterile centrifuge tube (Corning, Falcon) as it drips from the bottom. Alternatively, the virus band can be removed by side-puncture of the tube at the level of the virus band with a syringe and withdrawing the band with the syringe.

3. Dilute virus 5- to 10-fold in TSG (*see* **Heading 2.**, **item 17**); this will usually result in a stock that is 10^{10}–10^{11} PFU/mL when using wild-type adenoviruses.

4. Aliquot in 1- to 3-mL volumes in sterile 6-mL snap-cap polypropylene tubes or in cryovials and store at –70°C until needed.

5. Alternatively, if virus is to be used in animal studies (or in other experiments in which CsCl should be removed), the virus can be dialyzed against sucrose dialysis buffer (*see* **Heading 2.**, **item 18**, and **Note 11**).

6. Determine titer of the virus by plaque assay on A549 cells (*see* **Note 12**).

3.3. Plaque Assays on A549 Cells for Determination of Adenovirus Titers (see Notes 13 and 14)

One day prior to plaque assay, A549 cells are plated at 2.0×10^6 cells/ 60-mm dish (Corning, Falcon).

1. On the day of the plaque assay, dishes of confluent A549 cells are washed with 5 mL of serum-free DMEM for 30–60 min prior to addition of the diluted virus; this wash medium is removed immediately before the addition of the virus dilutions.

2. Serial dilutions of virus are made in serum-free DMEM; dilutions are done within a laminar flow hood. Virus is diluted in sterile disposable snap-cap polypropylene tubes, and tubes are vortexed well after each dilution (5–10 s at an 8–9 setting on a 1–10 scale). Typically for CsCl-banded stocks, the initial two dilutions are 1:1000 (10 µL into 10 mL), followed by dilutions of 1:10. Be sure to change micropipet tips after each dilution; avoid contamination of the micropipet barrel (use of barrier tips will help avoid contamination). Care should also be taken to avoid transfer of virus stock on the outside of the micropipet tip by avoiding dipping the tip into the solution, especially in expelling the volume. For CsCl-banded stocks, the range of dilutions that are usually countable is 10^{-8} to 10^{-10}.

3. A volume of 0.5 mL of the appropriate dilution is placed on confluent A549 cells (each relevant dilution is assayed in triplicate).

4. Dishes are rocked to distribute medium over the monolayer at 10- to 15-min intervals; cells are incubated at 37°C with 6% CO_2.

5. During this incubation, microwave the 1.8% Noble agar stock and place this in a 56°C water bath (at least 30 min before overlay). The 2X DMEM (*see* **Note 15**) should be warmed to 37°C, and other components of the overlay should be brought to room temperature before mixing.

6. Immediately prior to addition of overlay to the cell monolayers, mix the agar stock with the remaining ingredients.

7. At the end of the 1-h adsorption period, 6 mL of overlay (*see* **Heading 2.**, **item 23**) is added to the edge of the dish and the dish is rotated to blend the overlay with the medium used for infection (the 0.5-mL volume of medium used for infection is not removed).

8. Dishes are left at room temperature on a level surface for 5–15 min in order to allow the overlay to solidify.

9. Dishes are then transferred to 37°C and 6% CO_2 and incubated for 4–5 d. At that time a second overlay (5-mL/60-mm dish) containing neutral red is added (*see* **Heading 2., item 23**). It is important to have a humidified atmosphere in the incubator, but avoid very high humidity because it may cause excess moisture on and around the overlay, resulting in plaques that diffuse excessively and inconsistently.

10. Plaques are first counted 1 d after the neutral red overlay is added (*see* **Note 16**). On A549 cells it is not necessary to add more than the first and second overlays.

11. Plaques can be counted at 2- to 3-d intervals until new plaques are no longer becoming apparent. For Ad2 and Ad5 wild-type viruses this may be 12–15 d postinfection; with other serotypes *(5,6)* and with group C adenoviruses, which have mutations or deletions in the adenovirus death protein (ADP; previously called E3-11.6K), this may be approx 30 d postinfection *(7,8)*. Plaques are most apparent when holding the dish up toward a light source and observing an unstained circular area that has altered light diffraction. Cells initially may not be rounded up and may simply appear unstained, but plaques will typically become more apparent with time.

12. Dishes with 20–100 plaques are used for the calculation of titer (plaque assays are done in triplicate for each of the serial dilutions for more accurate numbers).

13. It is possible to analyze the kinetics of plaque formation to evaluate the virus's ability to lyse the cells at the end of the infectious cycle. To do this, count plaques from the first day that plaques are apparent on any dish/any virus. Do additional counts at 1- to 2-d intervals for all viruses. At the end of the plaque assay (when no additional plaques are becoming apparent [this may be nearly 1 mo postinfection]) calculate the percentage of the final plaque number apparent at each time point during the assay (*see* **Fig. 1**).

3.4. Plaque Assays on 293 Cells for Determination of Adenovirus Titers

Plaque assays are done in a similar manner to that for A549, but with the following modifications for the 293 cells:

1. Virus dilutions for the plaque assay infections are done in DMEM containing 2% FBS.

2. Do not rinse the monolayers prior to infection because 293 are very likely to detach from the surface. Care should also be taken to avoid excessive shaking or cooling of the 293 monolayers to avoid rounding up or detachment.

3. 293 cells are at 50–70% confluency when used for plaque assays (*see* **Notes 17** and **18**).

4. The infection medium is removed from the monolayer before the overlay is added. The overlay is added very carefully to avoid tearing the monolayer.

5. A third overlay may be needed if the plaque assays become more "acidic" (indicated by very yellow–orange color of the overlay). It is usually better to have the CO_2 level at 5% in the incubator in which 293 plaque assay dishes are being held.

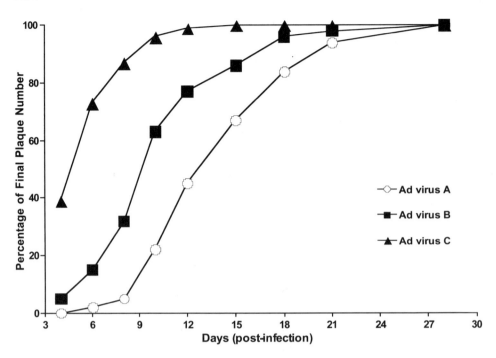

Fig. 1. Kinetics of plaque formation determined from plaque counts on sequential days of a plaque assay. The total number of plaques for a given virus on a particular day postinfection is shown as a percentage of the final plaque number for that virus. Viruses that are more cytolytic will show earlier plaque development.

4. Notes

1. Test bottles of medium are incubated at 37°C for 5 d to ensure sterility for each batch. Sterility can also be tested on blood agar plates or in nutrient broth. Care should be taken to cover the medium during preparation and to prepare medium in an area not normally used for handling of virus (adenovirus virions can pass through a 0.22-μm filter). Bottles, a 20-L container, and stir bar used for media preparation should be dedicated for tissue culture use and not mixed with chemical glassware.

2. Serum testing: sera (Hyclone, BioWhittaker, Invitrogen, or other suppliers) are purchased in large lots to reduce experimental variation due to differences in serum lots. Sera are tested with relevant cell lines in two different assays.

    a. In order to determine the cloning efficiency of cells at low cell density, 100–500 cells are plated per tissue culture plate. After 10–12 d, clones are fixed by methanol (10 min at –20°C) and stained with Giemsa stock staining solution or are fixed and stained with crystal violet (*see* **Heading 2.**, **item 27**). The number, size, and morphology of clones can then be compared for different lots. Determination of cloning efficiency is of importance for experiments in which small numbers of cells will be present on a tissue culture dish (as in production of stable transfectants or in limiting dilution for selection of clonal cell populations). One can also check the appearance of clones and make subjective judgments about the growth of the cells (flatness or overgrowth of the clones, size of the cells, cell uniformity, vacuoles, mitotic index, and relative "health" of the cells).

   b. Cell growth rates are calculated by plating 10^5 cells per 60-mm tissue culture dish; cell counts are done at daily intervals to determine the relative cell growth in medium containing the different serum lots.

3. FBS is not heat-inactivated (heat-inactivation will substantially decrease the survival of A549 cells in plaque assays).

4. Addition of neutral red to the first overlay may inhibit plaque formation and expansion. Do not reheat or reuse the overlay mixture.

5. We currently use Florence boiling flasks for our spinner culture cells because of their low cost and ease of cleaning. The flask is cleaned, the dumbbell stir bar is added along with ddH_2O, the flask neck is wrapped with a double layer of aluminum foil, and the flask is autoclaved before use. The stopper is wrapped in foil prior to autoclaving. Before use, water is removed from the flask, and the stopper is carefully inserted (the foil is unwrapped sufficiently to insert the stopper while maintaining sterility). The double layer of foil used to cover the top of the flask during autoclaving is now placed over the stopper, and the stopper can be removed while holding this double layer of foil.

6. HeLa spinner cells can be substituted for KB spinner cells. HeLa cells are typically maintained at a somewhat higher cell density and may require calf serum rather than HS for growth. The kinetics for virus growth would also need to be optimized to determine whether the 40- to 46-h time frame would be appropriate.

7. Higher passage 293 will work better for spinner cultures. Ideally, 293 cells should be at passage 60 or higher (but not passages higher than 100 passages).

8. Our experience would indicate that it is not necessary to do infections in serum-free medium or to totally replace the original medium on the cells.

9. Cells will typically appear enlarged, and often the nucleus will appear "blue" under phase-contrast microscopy when the cells are well infected and ready to be harvested. Cells will also be very round. Typically, dying cells will show distorted shapes; dead cells will look gray and "ghost-like." You will want to harvest the cells when most are rounded and enlarged, but before there is very much death. Under these conditions, essentially all of the virus will be within the infected cells, and the supernatants can simply be discarded.

10. Alternatively, virus can be extracted from cells using deoxycholate and Benzonase®. After the first PBS wash (**Subheading 3.2., d 4, step 3**), pellet the cells, remove the supernatant, and freeze the cells (–70°C) or process immediately. Resuspend the cell pellet (circa 10 mL) in 10 mL of 0.05 M Tris-HCl, pH 8.0, 1 mM $MgCl_2$. Add one-tenth volume (2 mL) of 5% sodium deoxycholate in ddH_20. Mix for about 10 min (the solution will become very viscous). Add 250 U of benzonase (Novagen, cat. no. 70664-3). Incubate at 37°C for 1 h. The

viscosity will be noticeably reduced after this step. Centrifuge for 20 min at 2100*g*. Collect the supernatant and discard the pellet. Add CsCl and band by ultracentrifugation as described (**Subheading 3.2., d 4, step 10** through **Subheading 3.2., d 5, step 2**). Collect the virus band, dilute in 0.05 *M* Tris-HCl, pH 8.0, containing 0.51 g of CsCl per mL. The virus is CsCl-banded a second time by overnight ultracentrifugation. Collect the virus band, dialyze against 0.02 *M* Tris-HCl, pH 8.0, dilute as needed, aliquot, and store in a buffer of choice (e.g., 0.02 *M* Tris-HCl, pH 8.0, plus 10% glycerol) −70°C. This extraction will give virus yields that are about three times higher than yields following freeze–thawing and sonication; the PFU:particle ratio is similar. The extraction works well for preparation of virus for use in infections. Traces of Benzonase might be problematic for extraction of viral DNA from these preparations. However, we were able to successfully extract viral DNA from these preparations by treatment with Pronase E (Sigma-Aldrich, St. Louis) at 1 mg/mL for 30 min at 37°C to remove remaining traces of Benzonase and subsequent disruption of virions in 1% SDS.

11. The banded virus is collected, inserted into a dialysis cartridge (Pierce), and dialyzed twice (in 2 L of dialysis buffer for 1–1.5 h each) at 4°C. The virus is then aliquoted and frozen at −80°C.

12. Verification of virus stocks: virus preparations are tested periodically by Hirt assay or other assays (such as immunofluorescence or PCR) to confirm the "fidelity" of the virus preparations. It may be necessary to plaque-purify a stock periodically to eliminate possible contaminants (this is especially important for viruses that grow less efficiently than wild-type virus or are released from cells less efficiently during infection).

13. Plaque assay consistency: careful and consistent plaque assays will typically result in less than twofold differences in determined titer of the same virus preparation in separate experiments. It is often preferable to plaque-assay a large panel of mutants that will be used in the same experiments simultaneously so that the relative titers will be quite accurate.

14. Adjusting plaque assays for small plaque morphologies (viruses that lack the subgroup C ADP gene and serotypes that produce small plaques): it is important to have consistent PFU information in order to do infections with viruses in which comparisons are made between the phenotypes of various virus mutants. The most direct method of generating infectious titers is by doing plaque assays for PFU. In the study of E3 mutants, we have determined that given mutants have a "small plaque" morphology, and therefore a number of modifications have been made in the previous plaque assay methodology to accommodate the requirements for these mutants. It is necessary for the cell monolayers to remain viable for an extended time (approx 28–30 d) to see the full extent of the plaque development. Plaque assays are done on A549 cells (ATCC), which have a very good survival time under the overlay in plaque assays. Cell survival of A549 cells is also somewhat dependent on the type of tissue culture dish used.

15. Care should be taken to avoid increases in pH (cell survival of monolayers in plaque assays is significantly decreased in medium that has become "basic" or if the pH of the overlay is too high initially).

16. For viruses that are more cytolytic or when the kinetics of plaque formation are being analyzed, the second overlay can be added as early as 3 d postinfection. Different colors of marker pens can be used on subsequent counts to distinguish which plaques were visible at which time point (this facilitates determination of the kinetics of plaque development).

17. Cell viability and plaque formation of 293 cells will be better if cells are infected at lower confluency. The monolayers are also less likely to tear or detach if not confluent.

18. We try to keep the 293 plaque assays as long as possible (preferably for close to a month). Plaques can be counted at 2- to 4-d intervals until no more plaques are becoming visible.

References

1. Pinteric, L. and Taylor, J. (1962) The lowered drop method for the preparation of specimens of partially purified virus lysates for quantitative electron micrographic analysis. *Virology* **18,** 359–371.
2. Mittereder, N., March, K. L., and Trapnell, B. C. (1996) Evaluation of the concentration and bioactivity of adenovirus vectors for gene therapy. *J. Virol.* **70,** 7498–7509.
3. Green, M. and Pina, M. (1963) Biochemical studies on adenovirus multiplication. IV. Isolation, purification, and chemical analysis of adenovirus. *Virology* **20,** 199–207.
4. Green, M. and Wold, W. S. M. (1979) Human adenoviruses: growth, purification, and transfection assay. *Methods Enzymol.* **58,** 425–435.
5. Hashimoto, S., Sakakibara, N., Kumai, H., et al. (1991) Fastidious human adenovirus type 40 can propagate efficiently and produce plaques on a human cell line, A549, derived from lung carcinoma. *J. Virol.* **65,** 2429–2435.
6. Green, M., Pina, M., and Kimes, R. C. (1967) Biochemical studies on adenovirus multiplication. XII. Plaquing efficiencies of purified human adenoviruses. Discussion and preliminary reports. *Virology* **31,** 562–565.
7. Tollefson, A. E., Scaria, A., Herminston, T. W., Ryerse, J. S., Wold, L. J., and Wold, W. S. M. (1996) The adenovirus death protein (E3-11.6K) is required at very late stages of infection for efficient cell lysis and release of adenovirus from infected cells. *J. Virol.* **70,** 2296–2308.
8. Tollefson, A. E., Ryerse, J. S., Scaria, A, Hermiston, T. W., and Wold, W. S. M. (1996) The E3-11.6kDa adenovirus death protein (ADP) is required for efficient cell death: characterization of cells infected with *adp* mutants. *Virology* **220,** 152–162.

Index